FIVE INNOVATIONS THAT CHANGE

We live in an era of major technological dev
social adjustment and dramatic climate change arising from human activity. Considering these phenomena within the long span of human history, we might ask: which innovations brought about truly significant and long-lasting transformations?

Drawing on both historical sources and archaeological discoveries, Robin Derricourt explores the origins and earliest development of five major achievements in our deep history, and their impacts on multiple aspects of human lives. The topics presented are the taming and control of fire, the domestication of the horse and its later association with the wheeled vehicle, the invention of writing in early civilisations, the creation of the printing press and the printed book and the revolution of wireless communication with the harnessing of radio waves. Written in an engaging and accessible style, Derricourt's survey of key innovations makes us consider what we mean by long-term change, and how the modern world fits into the human story.

Robin Derricourt is an Honorary Professor of History at the University of New South Wales, Sydney and a Fellow of the Australian Academy of the Humanities. His previous books in archaeology and history include *Inventing Africa*, *Antiquity Imagined*, *Creating God: The birth and growth of major religions*, and *Unearthing Childhood: Young lives in prehistory*, which won the 2019 PROSE Award in Ancient History and Archaeology.

Five Innovations That Changed Human History

Transitions and Impacts

ROBIN DERRICOURT
University of New South Wales

CAMBRIDGE
UNIVERSITY PRESS

CAMBRIDGE UNIVERSITY PRESS

Shaftesbury Road, Cambridge CB2 8EA, United Kingdom

One Liberty Plaza, 20th Floor, New York, NY 10006, USA

477 Williamstown Road, Port Melbourne, VIC 3207, Australia

314–321, 3rd Floor, Plot 3, Splendor Forum, Jasola District Centre, New Delhi – 110025, India

103 Penang Road, #05–06/07, Visioncrest Commercial, Singapore 238467

Cambridge University Press is part of Cambridge University Press & Assessment, a department of the University of Cambridge.

We share the University's mission to contribute to society through the pursuit of education, learning and research at the highest international levels of excellence.

www.cambridge.org
Information on this title: www.cambridge.org/9781009523370

DOI: 10.1017/9781009523400

© Robin Derricourt 2024

This publication is in copyright. Subject to statutory exception and to the provisions of relevant collective licensing agreements, no reproduction of any part may take place without the written permission of Cambridge University Press & Assessment.

When citing this work, please include a reference to the DOI 10.1017/9781009523400

First published 2024

Printed in the United Kingdom by CPI Group Ltd, Croydon, CR0 4YY 2024

A catalogue record for this publication is available from the British Library.

A Cataloging-in-Publication data record for this book is available from the Library of Congress

ISBN 978-1-009-52337-0 Hardback
ISBN 978-1-009-52339-4 Paperback

Cambridge University Press & Assessment has no responsibility for the persistence or accuracy of URLs for external or third-party internet websites referred to in this publication and does not guarantee that any content on such websites is, or will remain, accurate or appropriate.

Contents

List of Illustrations   page ix

1 **Introduction: Transitions and Impacts**   1
2 **Taming Fire** 10
   The Discovery of Fire 10
      Using Natural Fire 13
      Did Neanderthals Control Fire 16
      Homo sapiens and the Taming of Fire 18
   The Transformative Power of Fire 22
      Light 23
      Signalling: Home Base and Conflict 24
      Heat 27
      Social Impacts 28
      Cooking 29
      Biology 31
      Food Acquisition 33
      Technology 36
      Ritual and Art 38
      Myths of the Origins of Fire 41
   Conclusion 42
3 **Domesticating Horses** 44
   Histories of the Horse 44
      Horses before Domestication 47
      Horse Biology and Domestication 49
      The Horse before the Cart: Botai and the First Taming . . 51
      Domesticated Horses in the Western Steppe 54
      The Cart before the Horse: Animal Haulage by Ox (or Donkey) 56
      Horses and Chariots 61
      The Spread of Riding 66

	And to the East?	67
	The Multiple Roles of Horses	69
	The Horse for Meat	70
	The Horse for Milk	71
	Secondary Products	71
	Riding to Hunt	72
	Riding to Herd	73
	Riding to Communicate	74
	Riding to Fight	76
	Horse and Chariot	79
	Racing the Horse and Chariot	81
	The Pack Horse	81
	The Horse for Haulage of Goods	82
	The Horse for Haulage of People	84
	Riding to Travel	84
	Horses in Ritual and Symbolism	86
	Conclusion	87
4	**Developing Writing**	**89**
	The Origins of Writing	89
	Symbols before Writing	93
	Writing Numbers	95
	Pictograms, Ideograms and First Writing	98
	The Development of Cuneiform	100
	Hieroglyphic and Hieratic	102
	The Idea of Writing	107
	Other Independent Origins	108
	The Early Impacts and Roles of Writing	111
	Ownership, Identity and Trade	111
	Land, Rent and Building	112
	Administration and Authority	114
	Education and Lexical Reference	117
	Literature	119
	Social Class, Social Role and New Careers	121
	Propaganda and Pride	122
	Unifying the State or Nation	127
	Religion and Words to the Gods	127
	Spells and Magic	128
	Law and Political Regulations	130
	Correspondence	130
	Memory	131
	Conclusion	133

Contents vii

5 Inventing Printing 136
The Development of Printing and the Book 136
Printing in China 138
Printing in Korea and Japan 142
Invention of Paper 143
Scribal Copies 145
Precedents for Printing in Europe 147
The European Printing Revolution 147
Gutenberg and Mainz 151
The Spread of Printing 154
The Impacts and Context of the Printed Book 157
The Book Itself 158
The New Occupations 160
The Reading Classes 162
Language and Identity 166
Genres 168
Education 170
Intellectual Enquiry, Renaissance Humanism and Scientific Development 173
Ignorance 177
Public Discourse and Propaganda 178
Reformation: The Religious Revolution 179
Anxiety about Printing 183
Conclusion 185

6 Communicating Wirelessly 187
From Unwired to Wired to Wireless 187
Foot Transport and Horse Messengers 189
Ships and Colonial Empires 191
Semaphore 193
Pigeons 195
Telegraphy 196
The Wired Telephone 200
Communication without Wires 201
The Changed Wireless World 206
Naval Communication 206
Marine Safety and Weather Information 209
Military Tool 210
Direction Finding and Navigation 213
Radar 214
Colonies and Dominions 215
News and Propaganda 216
Entertainment Broadcasting 218

	Television	221
	National Unity in Language and Culture	223
	Connecting the World	225
	Criminal Apprehension	226
	Financial Operations	226
	New Careers	228
	Low-Power Devices	230
	The Internet: Cables and Wi-Fi	231
	Radio Telephony	231
	Conclusion	233
7	**Innovation, Progress and Presentism**	**235**
	Innovation	236
	Progress	242
	Presentism	246

Notes — 251

Further Reading — 280

Index — 286

Illustrations

1. Kalahari San making fire with friction. . . . . *page* 21
2. Lamps of the European Upper Palaeolithic. 25
3. Palaeolithic hearth: a diorama at Şanlıurfa Museum. 27
4. Rock painting of a figure smoking a beehive, Toghwana Dam, Zimbabwe. 34
5. 'Aborigines using fire to hunt kangaroos', Joseph Lycett, ca. 1817. 35
6. Palaeolithic artists in Font-de-Gaume, Charles R. Knight, 1920. 39
7. Upper Palaeolithic cave painting of a horse at Lascaux. 49
8. Palaeolithic horse sculpture from Vogelherd. 50
9. Wagon from the Royal Standard of Ur. 59
10. Assyrian king Ashurbanipal hunting on horseback. 73
11. Assyrian archers. 77
12. Mounted archer from the Mongol army. 78
13. Pharaoh Ramesses II and his chariot. 80
14. Covered carriage with women passengers, 1411, from the Toggenburg Bible. 85
15. Sumerian numerals in cuneiform. 96
16. Egyptian hieroglyphic numerals. 97
17. The development of cuneiform writing. 101
18. Umm el-Qa'ab tags. 105
19. Oracle bone. 109
20. Stone stele of Ušumgal, ca. 2750 BC, marking real-estate transfer. 113
21. Sumerian clay tablet with administrative account of barley distribution. 116
22. Ebony label from Abydos with the name of King Aha. 125
23. Panel from the tomb of Hesy-Re. 126
24. Chinese printing from the Song dynasty, 960–1269. 140
25. Sixteenth-century German printing press. 150

26	Typeface of the Subiaco Lactantius (1465).	152
27	The Gutenberg Bible.	153
28	Publication about Amerigo Vespucci's travels to the New World, 1504–1505.	156
29	A *Gest of Robyn Hode*: ballad from ca. 1510.	164
30	Textbooks in use: a woodcut from Perotti, *Rudimenta grammatices*, 1495.	172
31	Pony Express stamp, 1861.	191
32	Semaphore signalling at Scheveningen, 1799.	194
33	Goliath steamer laying telegraph cables, 1850.	197
34	Marconi in 1896 with his experimental radio equipment.	202
35	Reginald Fessenden working on radio voice transmission.	204
36	A German field radio station in the First World War.	212
37	Aeriola Sr Receiver, RCA Westinghouse, 1921–22 radio.	219
38	BBC Television camera with cable link to Alexandra Palace, 1938.	222
39	Wireless operator recruitment during the First World War.	228
40	Radio assembly work, Philadelphia, 1925.	229
41	Acheulean bifacial flint handaxes.	241

CHAPTER 1

Introduction
Transitions and Impacts

What innovations marked major transitions in the long trajectory of human history? What impacts did these bring about?

In this book, I present five key developments in human history which resulted in dramatic, widespread and long-term changes to the way we live. I describe what we know (from archaeology and history) of the origins and early development of each innovation and suggest what were their multiple and diverse effects.

While there have been other discoveries, inventions and changes which altered a single aspect of human life, the topics chosen for this book facilitated many transitions in society. Indeed, innovations whose origins lie in the deep past may continue to affect daily life in modern times. The subjects featured here are the taming of fire, the domestication of the horse (and development of the wheeled vehicle), the invention of writing, the technology and use of the printing press and the revolution in wireless communication.

Each of these developments brought major benefits to those who possessed the new skills, in contrast to those who lacked them. The story of an innovation is also the story of its spread, across time and space. A new capability can establish a new level of control over the environment, as did the ability to create fire. It can change hunters' relationship with their prey, move people at speed or enhance an army's power in battle, as did horse riding. It can transport goods in bulk and allow trade over new distances, as did the horse when combined with the wheeled vehicle. New means of communication gave power to those who mastered those skills, a feature of the beginnings of writing and the development of the printing press and printed book in Europe. Innovations can integrate widely separated groups of people, a function of the written word and the printed word and especially the era of the wireless revolution.

I suggest the topics selected for discussion in this book are distinct from many of the numerous important innovations which mark human history (and prehistory) because of the range and scale of the impacts they made. There are many kinds of innovation – technological, economic, political and ideological – and the final chapter in this book comments on some of these.

There were certainly significant ancillary impacts, for example, from the discoveries in the prehistoric world that effective composite tools could be made from small-scale stone artefacts, or that wild cattle could be domesticated for milk as well as for their meat and hides or that copper and tin could be combined to make bronze for tools and weapons. Even in the deep past, a combination of chance, creativity, experimentation and experience can be credited with the beginnings of a major technological or economic transformation in human culture.

We have become used to the concept of an invention or a discovery in modern times. An innovation such as penicillin or the flying machine can be adopted quickly and widely because of its power and value. The introduction of a new, basic foodstuff can have broad implications, including new settlement patterns and new trading networks. We have come to see that the 'agricultural revolutions' of the Old World and New World were slower and more complex than once assumed. Animals which were hunted prey were gradually domesticated. Control of crops brought about closer settled life, although agriculture could also lead to negative impacts on diet and to the rise of communicable diseases. The long and successful continuity of Indigenous societies in Australia affirms the flexibility and adaptability of foraging economies until the arrival of external agriculturalists was to undermine their long-lasting life patterns. The contact of New World with Old World meant the introduction of new foodstuffs in both directions.

Political change, power shifts, invasions and conquests are conventional markers in historical narratives. Many of these events served as the agents for wider changes in economic life and material culture. The Spanish destruction of Inca power in South America is an obvious example. We might cite the impacts of the Macedonian defeat of the Persian Achaemenid Empire in 334–330 BC, the unification of the Chinese state in 221 BC, or the fall of Constantinople to the Ottoman Turks in 1453. But many other changes of rule were merely changes of ruler. Even the rapid conquest of Byzantine and Sasanian territories by Muslim armies from Arabia in the 7th century AD had little immediate impact on either economy or material culture.

Introduction

Development and changes in ideologies impact more than just ideas; they stimulate political movements and population movements and create links between people which influence their material culture. So we should not underplay the power of religious or other idea systems, but their spread may well be related to other innovations, a topic mentioned in Chapter 5 on the printing revolution.

One feature of our perception of innovation is that the rate of change is increasing ever more rapidly. Is this because, in a world connected by instant communication and worldwide media, something new can spread so quickly, or is it the information about something new that is the fast traveller? How influenced are we by the sense that the present era is the inevitable climax of human history with the speed of new achievements unmatched in the past? This topic, too, is explored in the final chapter of this book. The present text was begun while the world was facing the COVID pandemic which, we were assured, was changing the world permanently. Debates about the impacts of human activity on climate change have gone further in marking our era as one of global environmental transition, not just adjustments in human society. The impact of environmental factors on the history of humankind and our earlier hominin relatives may be broader than any innovation brought about by human agency.

It is a commonplace, then, to note that there may be other side effects and impacts from any single development, invention, innovation or introduction, whether in technology, economic resources, ideology, political power or geographic settlement. But certain innovations, such as those discussed in this book, are transformative in the sense that their numerous and diverse impacts may cut across technology, economy, politics, ideology and geography. However, in presenting suggestions about their impacts, one must also be cautious in defining cause and effect. Correlation is very different from causation; it may be clear when a new development leads directly to another social change, while in other cases both may derive from the same source or have quite independent origins. Care in interpretation is always needed.

As these chapters show, there are stages in innovation. There may be marked steps: the use of wildfire available after lightning strikes was very different from the ability to create fire at will. The domestication of the horse and the development of the wheeled vehicle were separate processes, with greater and further impacts when the two were brought together. Writing was invented independently in different areas of the world and initially served different functions. Even printing had separate and probably unrelated origins in East Asia and Western Europe, although its spread in Europe was remarkably rapid as its commercial value matched market demands in linked regions. The wireless revolution took place in an increasingly unified world, where it enabled scientific communication to become ever faster and easier, but still the stages of development were distinct. We may have an image of the genius inventor, but as has often been stated, an invention is adopted and spreads only when it meets a contemporary or emerging need. Many 'discoveries' have had false starts, in a context not yet suited to their adoption, and our own experience reminds us of many apparently valuable innovations which prove to have no staying power as the world moves on or reasserts its traditions.

The spread of a new development may also be restricted, not only by geography, but also when a group may not wish to share it with another rival community. In competition for economic resources, and in the conflicts and wars that may result from that, the strength gained from an innovation (whether economic or material) can be so important as to make it essential to contain and restrict the knowledge for as long as possible. If knowledge is strength, maintaining the ignorance of others is also a strength. Fire could allow early *Homo sapiens* to dominate new territory in Europe and Asia through winters; cavalry could give a definitive military advantage over infantry; the written word gave power to the literate; and the printed text challenged those whose authority lay in just the spoken word. Wireless transmission of information is so powerful that it challenges authorities who may wish to control access in peace or war.

There is, inevitably, a personal and subjective choice in the topics I have selected for the present book. Each brought about transitions in human society, with impacts which foreshadowed longer-term development. Their individual importance is clear if we track back through time, imagining the past world without such additions to the armoury of human skills.

A modern world without wireless communication, or the technologies enabled by controlling radio waves, would deprive us of personal, social, work, economic and administrative activities. These appear daily essentials for the individual, the family, wider society, commerce, administration and the state. Yet until the 19th century (and in many world contexts well into the 20th century), all communication was limited by the proximity of humans to each other.

Before Europe's printing revolution of the 15th century, words were written individually one at a time in a single manuscript document. This meant limited access and allowed control of that access by the institutions of church and state. The remarkable speed with which the printing industry

Introduction

developed in the states of Europe showed the appeal of putting multiple and identical copies of a work, short or long, into the hands of readers throughout the continent. If printing was a handmaiden of the move from the medieval to the modern world, how different would have been historical change without that innovation in that place and at that time?

The invention of writing – symbols that represent sounds, rather than depicting objects, to those trained to read them – was equally revolutionary in its impacts, even though some of these impacts appeared after several centuries rather than in the initial phase. If we consider radio communication an essential marker of the post-industrial world and printing an equivalent for the post-medieval world, writing was a core feature of the post-agriculturalist world, the first urban socially stratified civilisations of the Middle East five millennia ago. Without writing, we can say that ownership and trade, commercial agreements and laws, political propaganda and submissions to the gods, religious teachings and poetic compositions were all limited to oral presentation and human memory.

Wireless communication, printing and writing are part of a sequence that begins with language itself: the relations between intelligibility and accessibility.[1] Language enables communication within a group but marks that group off from others. Writing divides society into the literate and non-literate. Printing widened access to the written world but favoured the growing literate classes, while wireless technology has dramatically broadened and changed (and in many ways democratised) the nature and control of communication.

The wheeled vehicle was known before the first civilisations and is still the basis of land transport of humans and goods. The horse was domesticated elsewhere to be ridden for hunting and herding. A tamed horse meant that human movement was no longer limited to human pace or water transport. When horses were combined with wheels on chariots, carts and wagons, transport itself was transformed. In time, horse-drawn vehicles and the individual on a horse were the basis of daily life, interaction and the economy in numerous societies. Horses have ceased to have such prominence today, but literature and historical records before the 20th century remind us of the dependence of so many Old World communities on our most important domestic animal, as well as the impact on its introduction into the New World.

The transformative power of fire is so dramatic that its impacts can be described as biological as much as historical. Fire changed where we could live and how we converted potential foods into energy and would form the focus of community lives. Its impacts went much further. We take for granted the ability to create, control and transport fire as part of the deep time of humanity, but that knowledge is only clearly associated with our own species of anatomically modern humans; our earlier ancestors may have gained chance benefits from capturing wildfire from vegetation lit by lightning strike.

Human society is bounded by time and space. Sociological theory has explored understandings and transitions in time/space, and the processes of restructuring and arguably compressing time/space in the modern era.[2] Historians mostly set their narratives in such a framework, while archaeological interpretations are typically framed within (and often defined by) chronological and geographical constrictions and questions of continuity and change. A feature of the subjects discussed in this book is their ability to expand or alter human relations with space and also time.

Radio communication cut across the limitation of face-to-face exchanges and the time delays in passing messages between groups. It developed so that information such as breaking news of a cyclone or the start (or end) of a conflict could spread rapidly and worldwide. Printing conveyed a single identical message in a book distributed across numerous state borders and available then to read and reread (or reprint) for decades and centuries. A written message could be sent across great distances without changing contents or relying on a courier's memory. A stone inscription could outlive the ruler who commissioned it; some forms of written words survived for millennia to the present day. The horse and the wheeled vehicle transformed the space accessible to a human group and reduced the time to travel between locations. The light from tamed fire could extend the day beyond sunlit hours, and its heat could maintain settlement areas through a season of winter cold. But in any historical development, 'human progress' is limited to, but importantly by, those with access.

In this book, I am concerned with the kind of impacts these innovations made early in their presence in human history. Of the topics selected, some impacts of their introduction may come more obviously to mind. Fire lit our nights, made us warm and cooked our food. Horses gave us speed and transport. Writing allowed the development of records and literature. Printing spread knowledge and debate, not least in the Protestant revolution. Wireless technology gave us the radios that link us to the wider world and has now given us the smartphone. But I argue that a single innovation could transform multiple aspects of our ancestors' lives and set new agendas for their future. There may be distinct stages in the early development and adoption of the innovations, each with their own impacts. Uses and further impacts continued to develop through time, with many adaptations,

Introduction

modifications and extensions, but this book is not a world history. The long sequence of roles and applications for each of these innovations can require, and indeed has generated, lengthy and detailed accounts. The emphasis here is on transition in human society, not the long subsequent experience.

The descriptions and discussions in this book draw their information both from historical studies and from archaeology, with questions inspired by other fields of knowledge. The narrative accounts have different characteristics the further back in time we look. The history of wireless technology and radio-wave applications is well enough established, although national pride puts different emphases to the fore in telling the story of remarkable discoveries, innovations and applications. The history of the 15th-century introduction of movable type and the printing press in Europe is also well enough established, with some disagreement on details relating to the life and specific authority of Johann Gutenberg himself. Scholars now recognise the earlier development of printing in East Asia, even if there is no evidence that the idea of printing spread from there to Europe.

The first writing is found in the earliest urban civilisations of the Middle East: Egypt, southern Mesopotamia (Iraq) and south-western Iran, with subsequent independent invention in China and Mesoamerica. Our knowledge of early writing in Egypt, in particular, is limited because of the friability and fragility of papyrus and linen as writing materials, in contrast to the fired clay tablets of western Asia. New archaeological finds of early writing may alter our interpretations. Meanwhile, questions remain about the timing and mutual influence of the idea of writing between these different contemporary emerging states in the Middle East. Note that the term 'Near East' is commonly used in the literature of archaeology and history to refer to the areas dominated by ancient civilisations in what today we call the Middle East; so both terms may be found in the text of this book and in sources cited.

The nature of the early societies we describe as civilisations is complex. They developed population concentrations in cities, the technique of writing, monumental architecture, centralised temple cults, state authorities and administrations, and social classes distinguished by wealth, role and power. The association of these features is open to differing interpretations on the nature and interrelationship of these different facets of the earliest civilisations, debates which go beyond the scope of this book.

In discussing the origins of wheeled vehicles and the domestication of the horse, we are reliant on archaeological work. Information in some areas is rich, such as evidence from wagons buried with humans; elsewhere, individual pieces of evidence of wheels can suggest an early date. The beginning of horse domestication is much more controversial, with well-argued and sometime passionate claims interpreting indirect evidence as proofs of human taming and control of horses. The picture has been enhanced by DNA studies, and new data and arguments are constantly emerging. The account in this book seeks to lay out some of the issues and suggests probabilities, but the detailed references in the notes are necessary to gain a fuller picture of the issue.

Similarly, there are diverse views in discussions on the origins of fire and the debate on when our human ancestors gained the ability to manufacture and control fire as and when required. Identifying evidence of fire within excavated sites occupied by prehistoric groups is relatively well established (despite some notable misinterpretations and some ambitious claims). But fire may occasionally have had an accidental presence in the site: it may have been brought in following a natural fire from a lightning strike, or it may be fire created artificially by friction or a stone strike-a-light. Here again, the discussion is amplified by the notes, which indicate some of the sources of debate.

Readers approaching these chapters on early innovations from the standpoint of archaeology may engage with such detailed accounts of current evidence and debates. Those with more historical interests may choose to focus on the accounts of their impacts and skim over the background chronological discussions on the taming of the horse and of fire, thinking perhaps that the more we find, or think we have found, the less we are sure we know.

This book has, I hope, avoided the simplistic view that human history is a one-directional (if interrupted) journey of progress with the present as its glorious pinnacle. As I have mentioned, the impacts of change can include positive benefits but may have other side effects; one person's gain may be another's deficit. Our society is not unique in seeing itself as the highest achievers in history, but the bias of 'presentism' can distort our perceptions of the past. I discuss this topic more fully in the final chapter of the book.

A linear narrative of history in the literature of today's wealthier nations often still places European (or European-American) civilisation at the core of the story of human achievement, stretching that Eurocentrism back in time. The studies in this book may help to inhibit such an approach. The taming of fire seems most probably to have emerged in Africa with our own species. Horses were first domesticated on the steppes of Central Asia. Writing was an invention of the Middle East, as well as China and Mesoamerica. Printing was first developed in China, Korea and Japan

Introduction

before it emerged in Western Europe. The control of radio waves was an innovation shared between the Americas, Western and Eastern Europe.

Each of the following chapters uses the same basic structure. I have outlined what we know, from archaeology and history, of the first stages of the development and adoption of the innovations discussed. I then suggest with examples their range of impacts. My acknowledgement of the many contributors to our knowledge and understanding is clear from the notes. Sources cited are necessarily fuller when addressing themes where the history is still less definitely established. Selected major works for each chapter are listed for further reading. The final chapter of the book addresses some broader issues of what we mean by innovation and progress and how much our perspectives are biased by our focus on the recent and our sense of the importance of our own species.

CHAPTER 2

Taming Fire

> Man in the rudest state in which he now exists is the most dominant animal that has ever appeared on this earth ... He has discovered the art of making fire, by which hard and stringy roots can be rendered digestible, and poisonous roots or herbs innocuous. This discovery of fire, probably the greatest ever made by man, excepting language, dates from before the dawn of history. These several inventions, by which man in the rudest state has become so pre-eminent, are the direct results of the development of his powers of observation, memory, curiosity, imagination, and reason.
>
> Charles Darwin, *The Descent of Man*, 1871

The Discovery of Fire

The taming of fire created dramatic and long-lasting transformations in human communities, with immense and varied impacts on personal, social and economic life. We take for granted warmth, light and cooking in the daily lives of societies we know from history or study in the archaeology of our own species. But the lifestyle of such communities required the ability to create fire whenever needed, maintain it, control its spread, move it within or even beyond the domestic setting.

Before the control of fire, our ancestors' days would have been shorter, their winters colder, their food raw and dangers many. Once humans had acquired the ability to create fire at will, control and transport it, their world changed. Fire transformed dark into light and cold into warmth and provided a focus for the camps of hunter-gatherer groups. It allowed game drives and management of landscapes to encourage browsing animals. Having cooked food as a regular part of the diet expanded available foodstuffs, removed toxins, required less energy to chew and digest and

The Discovery of Fire

provided more energy for the brain. Fire could be used to soften binding material, harden wood and prepare stone for making tools and weapons. And most striking to us, under artificial light advanced forager communities of the European Ice Age could enter dark caves and make vivid paintings on the cave walls. Myths and beliefs recognised the acquisition of fire as a major transformative event in human history; and from its Palaeolithic origins until its replacement by electrical power, fire was at the core of society.

The impacts that were achieved by human control of fire were many. Hominins had always hunted for meat and gathered plant foods; they had always sought shelter from the elements and migrated in response to climate as well as to availability of resources. The taming of fire – the ability to create it at will – established a new kind of control over the environment.

Without fire, prehistoric lives would be framed and limited by the changes in natural light set by dawn, dusk and the seasons; by the heat provided by the sun and natural shelter; and by what food could be eaten and digested in its raw form. Such was the experience of the ancestors and relatives of *Homo sapiens*. Temporary opportunities were available to them after lightning strikes if they were able to capture fire from burning vegetation and convey it for a period to their own home base. But then humans discovered the ability to create fire wherever and whenever required, by the use of wood friction (or alternatively creating a spark with stone percussion), to maintain it and even to move fire from place to place. Once the technology of creating fire was part of a society's cultural arsenal, it was unlikely to be forgotten or abandoned. It transformed life.

Language had developed gradually in prehistory. The technology of tools was developed with progressive refinements and evolution in the shaping of stone or wood. Fire once tamed offered up its multiple roles and means of social transition, a diversity of impacts which we explore in this chapter. The ability to manufacture fire at will was different from other developments of deep time – either you had it or you did not.

When did we achieve the ability to create fire at will and use its vast potential to change personal, social, economic, technological and non-material aspects of life? The dates, contexts and regions where such knowledge developed have been the subject of long and continuing debate among prehistorians. Claims have been made for evidence of early use of fire, and early manufacturing of fire, and then challenged, interrogated and tested with changing methodologies, leading to continuing debate rather than consensus.

Although we cannot identify a single place and time for the 'invention' of the human manufacture of fire, it is associated with anatomically modern humans as they developed in Africa. As *Homo sapiens* spread from there through the continents of the world, with a major impact from about 60,000 years ago, fire was a normal part of their culture and lives.

From its first taming by Palaeolithic hunter-gatherers, raw fire continued to be central and essential to human life until the introduction of domestic gas from the early 19th century (itself of course requiring fire to light it) and the generation of electrical power from the later 19th century. Both these processes took time to spread. Before then, lanterns and candles, hearths and forges, the first technologies of the industrial revolution and the destruction of enemy cities by fire all track back to the transformative point when fire was tamed.

The question is not a simple one of 'when fire?'. We need to distinguish the different relationships to fire of our ancestors and relatives we describe as hominins (to distinguish them from other great apes past and present). There is a passive response to naturally caused fire; and there is an active response using the opportunities provided by natural fire where it occurs. Then there is the ability to prolong a natural fire, by transferring it to a place of choice. Such use of fire implies carrying fire in a slow-burning material, still seen in modern contexts. Smouldering dry leaves packed inside a bark roll or hollow tree branch could be used to transfer naturally occurring fire to an area of occupation and could even be carried between places of temporary settlement. Embers carried inside an animal horn could serve a similar function.

But the final achievement was the *taming* of fire: the discovery and use of a technology to create then control a fire wherever and whenever needed. These then are different stages which parallel human evolution: adaptation to fire, use of fire and taming it to create, manage and apply fire in diverse contexts.

It is credible to suggest that once the ability to make fire had been established, it would not readily be abandoned. But that does not rule out its development independently in different places. Because fire is created most commonly from the heat of friction, the friction method may have been encountered when wooden boring tools were developed for other purposes, and their use may have stimulated the accidental discovery of friction generating heat to light a fire.

All these stages developed while humans and our hominin relatives and ancestors were foragers – hunter-gatherers moving across the landscape in response to economic opportunities afforded by the changing seasons.

The Discovery of Fire

Favourable occupation sites in the open or under the cover of rock shelters or cave entrances, and supplied with adequate and accessible fresh water, could be occupied briefly or serve as a base over an extended period; and a suitable site could be the destination for regular return occupation over an annual seasonal pattern. Settlement sites favoured by prehistoric hunter-gatherers (such as large rock shelters) can provide us with a record of successive occupation over tens and even hundreds of thousands of years.

The challenge is to identify signs of burning in the deposits recorded in the archaeology of such sites. These may be clear deposits of ash and charcoal, or the more indirect evidence of a level of burnt rock, or finds of burnt bone or the effects of burning on stone artefacts. Earlier excavations reported such finds based on visual examination and sometimes a little optimism. Interpretation of evidence of burning at some early sites was undermined when these proved to be manganese dioxide or organic matter or stained deposits, while others have been confirmed as the result of fire.

Today, a multitude of laboratory tests can assess the presence or absence of suspected burning. But in itself this does not demonstrate human agency, even when the burnt material is closely associated with evidence of human (or hominin) activity. A grass fire rapidly passing over a site of human occupation can generate sufficient concentrated heat to burn and transform the raw bones of a human meal or abandoned stone tools; and a burnt tree stump at a site of prehistoric human occupation can resemble the concentration of a deliberately created hearth.

When a constructed hearth is found associated with the food waste and tool debris of human settlement, we can be confident of its role in fire management. But observations of fires in modern and recent forager groups show how readily the signs of an active hearth may disappear; the absence of evidence of burning from a prehistoric site is no confirmation that its occupants did not know and use fire. Understanding when and where fire was used in prehistory involves careful assessment of evidence, context – and probability.

Using Natural Fire

The regions occupied by our earliest hominin ancestors included widespread African areas of grasslands extending below tree or bush canopy. The natural ecology of such areas is marked by the incidence of natural fires started by lightning strikes. The most frequent and dramatic of such fires are at the start of the wet season: after months without rain as the tropical heat dries out the vegetation, the first thunderstorms of the rainy season are marked by lightning strikes which ignite the dry grass.[1] Such fires move rapidly, generating their own wind currents as they develop, and as they pass through the grass undergrowth with great speed, they leave in their wake smouldering trunks of trees. Other sources of natural fires also exist: flows from volcanoes, even spontaneous combustion such as can occur in bat guano. There are reports of wind generating sufficient friction in bamboos or other tree branches to cause them to catch fire. But lightning fires and their aftermath are the most common likely source by which early hominins would encounter fire. The frequency of fires in a natural environment is substantial: the United States, for example, records over 30,000 fires a year, rural and urban, started by lightning, with an average of 2,000 fires a year just in the Northern Rockies.[2] Fire can make a difference in cold climates as a phenomenon of the warmer months: in Alaska, there may be 100 fires burning on any day in the summer season. Today, of course, most vegetation fires are caused by human action and often accidentally.

Studies by US anthropologist Jill Pruetz and others have shown the reaction to natural fires of non-human primates, noting how chimpanzees and monkeys can be aware of the dangers and anticipate the movement of a fire through the landscape. They avoid, predict and navigate around fires with apparent confidence and then forage for food in burnt areas.[3]

Wildfire has been present throughout hominin evolution, even if its location was unpredictable. The Australopithecines of Africa would experience grass fires lit by lightning, which spread across the savannahs and open woodlands which provided their foodstuffs.[4] They would have fled from threatening fires but also found new foods in fire-ravaged areas: plant foods burnt by the fires, lizards and animals who did not escape the fire, exposed burrows, birds' eggs and fire-burnt landscapes allowing new plant cover and attracting new browsers.

An interesting 'thought bubble' arose from the observation that Olduvai Gorge – home to a sequence of early hominins – had hot springs 1.7 million years ago that could have been used for transforming foods by boiling – not a readily testable proposition.[5]

As the *Homo* genus evolved in Africa and spread through the Middle East into Asia and Europe, we see scattered evidence of occasional use of fire.[6] *Homo erectus* (and their closest relatives) occupied parts of Africa alongside Australopithecines as early as 2 million years ago, as shown at Drimolen in South Africa, yet have been found as late as 120,000 years ago at Ngandong in Indonesia. It was *Homo erectus* who created and used the bifacial stone handaxe (the standard Acheulean equipment for over a million years in Africa, and

The Discovery of Fire

extending into the Middle East and Europe), while in much of Asia they were users of a different stone technology. Compared to Australopithecines, we assume their diet had a significantly greater contribution from meat.

Concentration in certain sites suggests that, from time to time, *Homo erectus* found themselves in the vicinity of natural fire and were able to capture it and bring it back as smouldering matter to use at their camp – whether for heat, to try its effects on the technology then in use or to transform foodstuffs. As stable and sophisticated hunters and food gatherers, *Homo erectus* would have known and used the effects of natural wildfire in driving game animals in an anticipated direction and scavenged where fire had killed animal life in its passage. As natural fire burned an area and stimulated new growth which in turn attracted grazing animals, the hunters would target the opportunities this provided. The infrequency of fire evidence implies that this was use of fire, not its control and management.

Evidence of fire in sites occupied by *Homo erectus* before about 400,000 years ago is sparse, and some such finds may reflect the chance passage of fire.[7] No more than a dozen locations in Europe and a dozen in Africa and the Middle East have evidence of burning within their human settlement areas. They include burnt stone artefacts at Koobi Fora in Kenya at a date of 1.5 million years and Wonderwerk Cave in South Africa at 1 million years. New techniques identified a slow burning process on animal remains in the Israeli site of Evron Quarry between 1 million and 800,000 years ago.[8] Two hearths in a Spanish Acheulean site dated ca. 245,000 years ago represent the earliest known use of fire in Europe.

The team who has studied the Acheulean site of Gesher Benot Ya'aqov in Israel has argued forcefully for the regular use of fire there around 780,000 years ago, and its regular control by the hominin occupants as an integral part of their culture. In addition to finds of burnt seeds and wood, concentrations of burnt small stone fragments were interpreted as 'phantom hearths'. Subsequent work located areas of fish teeth from large carp with no corresponding small bones, which the excavators argued must indicate the effects of cooking with controlled temperature: even, they suggested, the use of earth ovens.[9] These are ambitious claims.

One necessarily cautious observation is that the spread of the uses of natural fire suggests an ability to transmit cultural ideas about innovations and adaptations, even before the emergence of language.[10] We can see that material culture – not least in the style of stone tools – spread between groups; we can assume that patterns of settlement and use of the environment were shared; so copying uses of fire without the need for what we call language is a reasonable supposition.

A theory advanced by US evolutionary biologist Richard Wrangham, interesting and challenging but not widely accepted, argued that the use of fire was such a regular part of the lives of *Homo erectus* that it contributed significantly to the biology itself of their emergence as a species. In this model, called 'the cooking hypothesis', the regular use of fire in cooking and the consumption of food in cooked rather than raw form meant less energy went into chewing and digestion and more energy went into the development of the brain, thus enabling *Homo erectus* to make cultural advances. However, such a model would suggest more regular and earlier evidence of fire than has been found on archaeological sites. And if the larger brain of *Homo erectus* had developed because of a diet of cooked food, that would push the taming of fire back to the origins of that species rather than being a 'discovery' they made at some stage. Wrangham admitted this in suggesting that it would have been a group of earlier *Homo habilis* who acquired control of fire, allowing them to evolve into *Homo erectus* because of the advantages this gave them – a bold claim.[11]

Did Neanderthals Control Fire?

There is far more archaeological evidence of the use of fire associated with *Homo neanderthalensis* – the Neanderthals who occupied the southern and central regions of Europe and areas of the Middle East from about 400,000 years ago.[12] They were responsible for the more developed Middle Palaeolithic technology (making stone tools with a prepared core). After about 46,000 years ago, Neanderthals were gradually replaced by *Homo sapiens*, anatomically modern humans using an Upper Palaeolithic blade-tool technology who also interbred with Neanderthals in Europe.

There is increasing evidence of Neanderthal groups with social behaviours we had assumed were exclusive to our own species. Some Neanderthal societies buried their dead; some looked after the ill and injured; some had elements of personal adornment; even geometric shapes painted on cave walls in Spain have been assigned to them. Many scholars consider Neanderthals to have possessed a form of language which would aid in social organisation and planning. In this context, it is unsurprising that many now consider Neanderthals, rather than modern humans, to have been the first to tame fire and control it for their own use.

Excavations and studies of Neanderthal and Middle Palaeolithic sites well outnumber those of the Lower Palaeolithic and earlier *Homo*. Fire can be found in many Middle Palaeolithic locations of Europe's Neanderthals, especially after about 300,000 years ago, but it was by no means present in all. Charred seeds are found at some sites but may sometimes come from within dung burnt for fuel rather than representing foods. Burnt bones are more direct evidence of cooking meat. There is evidence for birch tar heated for use in hafting an implement.

Contemporary with the early European Neanderthals, fire seems to have been persistently used in occupations at Qesem Cave in Israel within (though not necessarily throughout) a long time period, from 400,000 to 200,000 years ago. Many ashy deposits contained burnt artefacts and burnt bones, and micro-charcoal was found in the dental calculus of the occupants' skeletons. Excavators at Qesem Cave argued that ash had been used to preserve food in even earlier periods.[13]

It seems clear that Neanderthals were willing and able to take advantage of fire when available. As with *Homo erectus*, the ability to capture natural fire, bring it to a home base and use it there for cooking and other beneficial purposes is clearly attested. That puts fire in the same category as other wild resources. But the ability to create fire wherever and whenever required is not yet proven for the Neanderthals.[14] We are not able to confirm it was domesticated to create at will and curated as part of basic Neanderthal technology and social norms. A study of use-wear marks on stone tools associated with later Neanderthal settlement in France suggested they were consistent with possible use to create sparks from pyrite to create fire, a process somewhat simpler than using a fire drill.[15]

The inconsistent and variable pattern of fire's presence at Neanderthal sites has stimulated much debate and theorising. One genetic study suggested that Neanderthals were more sensitive to the smoke from fire than modern humans, but this interpretation was rejected in a later genetic study. Another argument asked whether the effort of collecting firewood and starting a blaze may have outweighed the value of fire at a specific time and place.[16] But Neanderthals were not living in treeless sandy deserts or barren icy wastelands. When we consider how transformative is the ability to create and maintain fire whenever needed, and the substantial range of benefits provided to nomadic foragers by controlling fire, the idea that it was at times 'too much bother' sounds unrealistic.

One explanation advanced for Neanderthal periodic use of fire is that it helped them maintain settlement in the same areas despite serious declines in temperatures during the European Ice Age. The heat of fire would allow them to stay in economically valuable areas through the winter and to migrate and remain further north as population pressures required. But the picture is not quite so simple, for excavated sites in south-west France demonstrated greater evidence of fire use in warmer periods than in colder eras of the Ice Age. This could imply reliance on natural fire, since that would occur more readily when vegetation dried out by a hot climate was lit by lightning strikes.[17] Neanderthals' biological ability to withstand cold climates of the Ice Age is a feature widely commented on, although it was climate more than potential food resources that set the northern limit to their regions of settlement.

There is no substitute for fire in human society – or in hominin evolution. Stone tools can be replaced by wood or bone, animal foods can be replaced by plant foods, gestures and noises can be a stage towards the development of language, rock overhangs can serve instead of constructed shelters, but fire has no alternative, no equivalent. Once gained, it would be an essential part of human society. The irregularity of fire in Neanderthal settlements would support a view that they had not yet learned to tame it and create fire at will.

Homo sapiens *and the Taming of Fire*

We are more confident in associating the taming of fire with our own species, which had evolved within Africa after about 300,000 years ago. The site of Jebel Irhoud in Morocco with such early dates had fire-heated flint artefacts of Middle Stone Age type associated with fossil material considered to be early *Homo sapiens*.[18]

Knowledge of fire management accompanied our species in their spread from Africa after about 60,000 years ago: across Arabia and Asia into Australasia, into the Europe of the Neanderthals, then across into the Americas, and eventually to colonise the Pacific from Asia. We cannot identify one point in time and space where the creation and control of fire reached the new level of sophistication we call taming. Given the technological, economic and biological advances that mark the emergence of *Homo sapiens*, such a step in the management of fire could have been made in more than one location. Yet given all the advantages held by a mobile hunter-gatherer group (or indeed an individual) who could manufacture fire the rapid dissemination of the technology of fire taming and application from locations of origin would be unsurprising. In the major spread of modern humans from Africa, it is a reasonable hypothesis to suggest that they took with them a set of fire-control techniques first developed in the African continent.

The relations are uncertain between mastering the advantages of controlled fire, changes in stone-tool technology (the later Middle Stone Age of sub-Saharan Africa, the Upper Palaeolithic of Europe) and developments

The Discovery of Fire

in brain and cognitive development, language, symbolic behaviours, social interaction and community interdependence. Correlation is not causation. Modern humans have a smaller gut and lesser jaw than earlier hominins. Cooked food ensures less energy is required for chewing and for digestion and more energy is available for use by the human brain.

Fire taming means the knowledge of how to create fire wherever and whenever wished: 'habitual use'. The technology involves the ability to make fire across widely different environments, from tropical forests to arid tundra, most commonly by the friction of a hardwood on a softwood to generate a heat which will ignite kindling. Fire can also be made by striking a suitable stone (where available, flint) on mineral pyrites or marcasite. And also of course by capturing natural fire, as before. Maintaining fire in a constructed hearth topped up by firewood allows flexibility beyond the smouldering stump of a wildfire. Then carrying fire within and outside a campsite involves use of a char – vegetable matter tightly packed inside bark or horn, or smouldering fungus or dung.

Not every individual may have been equally skilled at creating a fire from friction. We can assume in prehistoric forager societies a preference to retain fires from day to day, but also often to carry it from place to place, rather than bring the appropriate hardwood, softwood, kindling and expertise for each place and time.

The location of a home-base campsite (whether in the open or in a rock shelter) is determined by the availability of fresh water at an acceptable distance. Access to firewood may be an influence, but it would be unusual to find a location that did not have burnable vegetation available for collection within the same range as plant foods. A very wide range of woods was considered suitable for Palaeolithic fires. At later prehistoric forager sites in the Weld Range of Australia, the species of wood used for fire were just those available in the areas visited to collect food and water, rather than more suitable woods from other locations.[19]

Features identified as hearths are a rarity in the Lower Palaeolithic and still uncommon in the Middle Palaeolithic. But a normal feature of occupation sites of the Upper Palaeolithic and of the later Middle Stone Age in Africa is a deliberately constructed hearth together with ash, burnt stone and burnt bones. Many of these hearths were stone-lined, and some were clay-lined basins with a clay-formed rim.[20]

Such hearths were often placed in the central area of a settlement to provide light and heat, although, as shown in the description of modern small-scale forager societies, a hearth may be in a marginal position to serve primarily as a place for cooking. A simulation study of the density and dispersion of smoke from a fire in a cave setting, compared to the hearths in the French Lower Palaeolithic site of Lazaret Cave, confirmed the unsurprising observation that the positions of the hearths there were chosen to minimise the residents' exposure to smoke. At the Nerja Cave in Spain, studies of soot as well as charcoal tracked occupation back to 41,000 years ago, the charcoal scattered from torches carried to light the dark interior.[21]

If taming fire involves its creation, maintenance, use and transport, what can we construct for the methods and approaches used? Direct evidence is rare and circumstantial; for guidance, we may have to consider observations of how people in more recent mobile small-scale societies managed fire. While like all of us these have the same biology and mental structure as the first anatomically modern humans, the different demands and opportunities of small-scale societies and economies dependent on food gathering, hunting and fishing help indicate how our ancestors may have managed the fire they controlled.

Methods of Making Fire

Fire can be created by striking a flint on pyrite, or by friction with a drill stick. Both are techniques which require planning and preparation. Friction has been the most widely available means used by mobile foragers to create a fire. A hardwood rotated at speed in the hollow created by a sharp stone awl in a softwood generates heat in the resultant wood dust; once applied to make dry vegetable matter smoulder or flame, this can be transferred to light the fire in a constructed hearth or open space. Modern bushcraft guides readily demonstrate such a process.[22] A single individual can use a cord drill to rotate the hardwood. The same effect can be efficiently achieved by using a bow with tough fibre to rotate the hardwood stick, a common tradition among northern Native Americans. The use of a loop of fibre to rotate a stick held by a second individual can aid the process. Alternately, a hardwood stick pushed along a groove in the softwood can generate sufficient heat to light the kindling.

A description of the process among the nomadic !Kung San Bushmen of the Kalahari may be little different from that used over the long human past (Figure 1). All !Kung men and women were competent in the fire-drill method, which involved ninety seconds of strong and controlled action.

> The sticks are about 50 cm long and 1 cm in diameter. Two different woods are used: a hard wood such as *Catophractes alexandri* for the drill and a softer wood such as *Ricinodendron rautanenii* (mongongo) for the

The Discovery of Fire

FIGURE 1 Kalahari San making fire with friction (photo credit: Jorge Fernandez/Alamy Stock Photo).

base. The operator cuts a notch near the tip of the base stick held flat on the ground, with a knife blade to receive the coal, and places the tip of the drill stick in the notch. He twirls the drill stick rapidly between his hands with a firm downward pressure, taking care to keep the drill tip from slipping out of the notch. After 10 seconds smoke begins to spiral up from the drill tip, and after 40 to 50 seconds a dribble of smoldering dust collects on the knife blade. After a minute or so, the glowing dust pile is carefully tapped into a wad of soft grass and blown into flames.[23]

The alternative and even quicker method from stone, for example by a 'strike-a-light' (striking a flint on pyrites to generate a spark which lights the kindling), also seems to have been used by prehistoric foragers. Evidence of this has been identified in the Upper Palaeolithic of Denmark and the Netherlands, but pyrites can readily disintegrate and disappear from an archaeological assemblage.[24] A rounded edge is produced by frequent use as a striker, and such a feature on flint artefacts occurs in sites of the period. Much later, flint and pyrites have been found together in European Neolithic and Bronze Age graves: fire-making equipment associated with the deceased and perhaps provided for use in an afterlife.

Once the kindling has been lit, maintenance of the fire requires the collection of adequate material. But this could be dried bone or dung as well as suitable woods.[25]

Hunter-gatherers observed in modern times commit only part of their day to collecting foodstuffs, but retrieval of material for a community fire is a demand of equal importance. Transporting fire from place to place requires a burning lamp, a smouldering log or a fire stick, to recreate fire at a new location or temporary overnight camp. Groups of Australian Aborigines would carry a 'glowing firestick' or bark torch while travelling.[26] Despite the importance of fire in the society of the Mbuti hunter-gatherers of the Congo, it is not remade for each use: highly effective means of carrying it wrapped in fire-resistant leaves allow fire to be moved from place to place without the need (or, it was said, the knowledge) for starting a new fire in each new location. A New Guinea group conceded that everyone knew how to make fire, but it was hard work: much better to borrow fire from a neighbour.[27]

The Transformative Power of Fire

The ability to use wildfire, on those occasions when it was available from nearby lightning strikes, is comparable to the harvesting of other natural resources by early prehistoric communities: wood and stone for tools, plants and honey and fish and animals for food. But quite different was the revolution in human life once the ability had been acquired to create and maintain fire whenever wished. That innovation was truly transformative, affecting so many aspects of social, family and individual life – one of the great changes in our cultural development. The domestication of fire, previously only known in the wild, could be compared to the much later

The Transformative Power of Fire

domestication of food animals and plants, yet its effects were even more dramatic.

While some of the transformative effects of fire are more immediately obvious, the social impacts and potential of the innovation are numerous, as outlined in the following sections. So much of what we assume as normal human life in the past depended on the ready ability to create fire, in contrast with the stages of earlier prehistory in which fire was an occasional bonus.

Accounts of the lifeways of forager groups of the modern era can help us understand the roles that fire can play and give us some understanding relating to earlier hunter-gatherer groups of our species. An ambitious study by archaeologist Lewis Binford used data from 390 different communities of modern-era foragers to illuminate issues relating to prehistoric hunter-gatherers. Another list identified twenty-seven groups of modern foragers, after excluding sedentary fishermen, those who hunted with the use of domestic animals and those who undertook some marginal element of farming.[28]

But it is important to remember that such studies were of technologically advanced hunter-gatherers; that the communities who survived as modern foragers were often living in marginal environments (dense jungle, arid semi-desert or northern regions of extreme cold). They often had client or exchange relationships with farming neighbours and by definition were in contact with the outside world at the time of published records.

Light

Without the light created by artificial means (fire carried in burning vegetation, lamps, candles and lanterns and later the illumination provided by gas and electricity), light is essentially available when provided by the sun. It may be supplemented by moonlight in the open air and cloudless settings, with a monthly cycle useful for some hunting expeditions after nocturnal animals. Thus, light defines the day and extends the time for human activity. The 7.5 hours of daylight for a Lower Palaeolithic forager in the midwinter of north-western Europe, for example, became a potential 24 hours for activity by the Upper Palaeolithic. The dramatic changes in daylight hours through the seasons of the year are reduced, and the patterns of economic and social activity and sleep are evened out. Choices of allocating time between food acquisition, social activity and rest become more flexible. Upper Palaeolithic lamps are typically made of limestone or sandstone with a hollow to retain a burning substance presumed to be selected animal fats. At the famous French painted cave of Lascaux, a carefully carved piece of red sandstone with a fine handle has been interpreted as a lamp to light the area for the artists, using animal fat with bark for a wick, although an early suggestion proposed it could serve to hold items for burning.[29] By the later Palaeolithic of around 20,000 years ago, some lamps were also decorated (Figure 2).

The light of a fire in a hearth can illuminate the evening for a whole community; the light of a lamp can allow people to be away from their home base at night. In an open campsite or a large rock shelter, a lamp facilitates safe movement away from the area of a central fire at any time.

Hunting at night with the use of mobile light can increase the range of available food resources – a notable advantage given the nocturnal habits of many animals. Without the availability of modern weapons, a traditional hunter would rarely expect a successful spear or arrow strike on a large animal to achieve an immediate kill, and the pursuit of an animal wounded or poisoned by a hit could take several days and nights, requiring fire for light, heat and cooking away from a home base camp.

The ability to create fire at will creates a functional space in a dark night. But what was the impact of this cultural change on human biology? Before the light of the campfire and the lamp, night activity was limited by the phase of the moon. Humankind was a diurnal creature adapted to the long patterns of sleep imposed by darkness. Biological rhythms were disrupted and modified by the regular use of light to extend the day and the social interactions of groups. Dusk is no longer the same; short winter days are not so short; the monthly cycle of the moon loses its impact. Studies of the dramatic modern impact of artificial light on multiple aspects of the lives of other species is a reminder of how the availability of night-time light may have affected our ancestors.[30]

Signalling: Home Base and Conflict

The presence of a fire at night provides a central focus for a family group and marks a symbol of the community and a focus for communal activity. Ethnographic studies have reaffirmed the significance of a central hearth, or distinct subsidiary hearths, in the lives and also the perceptions and ideology of small-scale mobile groups, extending beyond practical uses. Hearths are central to domestic, sub-group and larger community settings and can be seen as corresponding to

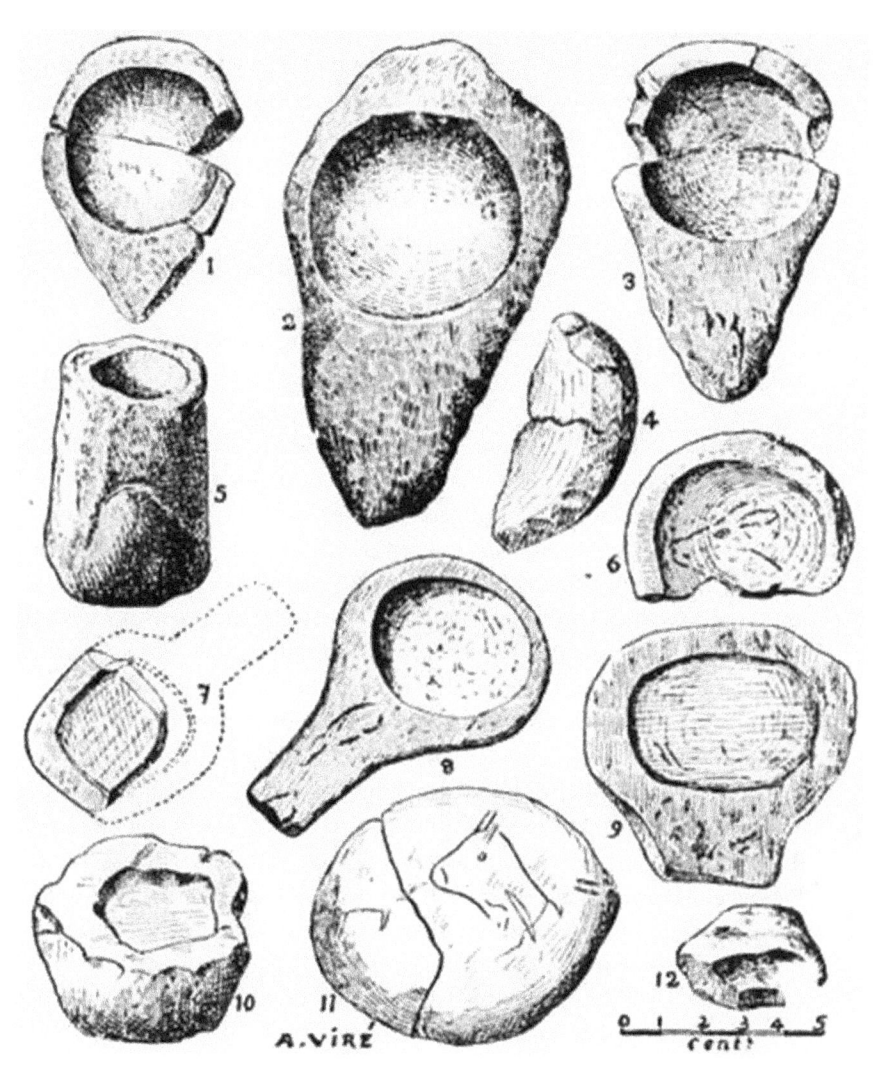

FIGURE 2 Lamps of the European Upper Palaeolithic (A. Viré, 'Les lampes du Quaternaire moyen et leur bibliographie', *Bulletin de la Société préhistorique française* 31 (1934): 517–520).

the complex pattern of social organisation. A new group camp requires a new central hearth created from the embers provided by each family in the band, and the new central hearth is then a place for adult males to gather and for ritual. The different hearths serve to bring children into the wider group at different ages.

When groups of foragers meet, the exchange and sharing of food provide an important signifier of the friendly purposes of such a meeting, and a central fire may support this goal. Studies of hunter-gatherers of the modern era suggest that individual mobile groups may have typically averaged around 25–30 people and may have formed part of a wider network of 500 or so, providing a context for finding marriage partners outside of closer kin. Where members of one group seek contact with another, the sight of a distant fire at night may serve as a guide to making contact in the day.

A hearth is also a definitive, even deliberate, marker of possession: a deserted site with a hearth showing evidence of recent occupation can be an indicator of territorial claims. Ethnographic studies serve as reminders that nomadic groups occupy and exploit territories that they claim as their own, while archaeology demonstrates the continuity of settlement of sites, subject to numerous return visits over a seasonal cycle.

No less important, a fire acts to keep predators away. Maintaining the light of a fire through the night and sleeping close to it is a sure way to deter opportunistic attacks from predators and discourage other creatures who may seek the same shelter. The security of a campfire provides reassurance from dangers, whether snakes or wolves, bears or leopards.

And while it may be premature to talk of 'military uses' for fire in the Palaeolithic, we know that inter-group conflict was a feature of society at least as far back as the Middle Palaeolithic. In the Upper Palaeolithic of Předmostí in the Czech Republic, multiple burials of adults and children imply a simultaneous violent death, and such violence is also seen in burials at Maszycka Cave in Poland. A cave fire would signal territory to warn a rival group, but such a group could use fire to destroy bedding and other materials in an enemy camp, pioneering the destructive use of fire in conflict which became the basis of so much devastation in the wars of 'advanced' societies of the agriculturalist world.

Studies of recent hunter-gatherer groups showed the use of fire away from the home base in a number of formal signalling roles. It brought together a raiding party, indicated the location of a hunting party or advised the approach of a herd of animals to hunt. In the frontier wars with European settlers in the 19th century, Australian Aboriginal hunter-gatherers were experts in using fire to signal between groups and assist in organising resistance. The Yahgan in Tierra del Fuego used smoke signals to advise of an accident or other emergency.[31] The ability to create fire when needed for any purpose takes cooperative human culture a major stage further.

The Transformative Power of Fire

Heat

The guarantee that a burning fire can be created whenever required transforms the daily and annual pattern of a forager group. The hearths of advanced hunter-gatherers may range from simple piled wood for short-term use to a construction of low stone barriers around a dug fire pit (Figure 3). The conventional image of Upper Palaeolithic life in Ice Age Europe is of a group, large or small, gathered around a significant fire, whose embers and charcoal might be used to make smaller fires within the same home base.

When night falls and the sun is gone, fire provides continuing heat through the evening and, of course, if kept fuelled, through the night and into the morning. An otherwise cold location in the open becomes a suitable camp site; a rock shelter or cave entrance remains habitable throughout the year. In winter, the !Kung San (Bushmen) of the Kalahari would stay around the campfire until mid-morning.[32]

And, perhaps more important, the permanent availability of fire meant that anatomically modern humans, whose bodies are less fitted to withstand the impacts of cold than has been suggested for Neanderthals, could

FIGURE 3 Palaeolithic hearth: a diorama at Şanlıurfa Museum (Dosseman, CC BY-SA 4.0, via Wikimedia Commons).

combat their biological limitations with artificial means (fire, but also fabricated clothing) to inhabit colder regions of Eurasia. Since humans migrated from cold regions of south-east Siberia into cold regions of northwest North America, we could credit control of fire as the agency which enabled the first populations of *Homo sapiens* to enter the Americas.

Physiology and climate determine the limitations of settlement regions and patterns of seasonal migration for all other species; but for *Homo sapiens*, these are augmented by culture – the ability to create and maintain fire gave our species opportunities outside of those environmental constraints.

In studies of modern forager groups, archaeologist Lewis Binford suggested a common model of sleeping between a series of fires which in the archaeological record would be represented by hearths relatively close to each other. Hearths thus mark out a personal space. Another survey notes fires in a rock shelter positioned along the inner walls of settlements both prehistoric and recent.[33]

Social Impacts

As any camping experience reminds us, a fire is more than a source for heat, light and cooking. It provides a central place for a community. Based on observations of forager societies of the modern era, we can imagine members of prehistoric bands involved in different combinations in different activities during the day.[34] The evening would be the regular context for the group to gather round the heat and light of the fire. Such social interaction provides the framework for the development of social planning – whether for daily food gathering or movement of the mobile group to another location.

In his study of the !Kung San of the western edge of the Kalahari, Richard Lee observed:

> The cooking fire (*da*) is a central element in the domestic, subsistence, and symbolic life of the !Kung. Around it, the !Kung spend most of their non-working hours. Every household has one; at it all the meals are cooked, and around it in the evening conversations take place and the household members spread their sleeping mats when they turn in for the night.[35]

A hearth was a centre of daily activity as well as the evening, as nuts were shelled, resin softened by the heat to haft a spear or a surface next to the hearth used for work.[36]

While some traditional forager groups observed in the modern era, such as the Hadza of northern Tanzania, had their own forms of social and recreational smoking – imbibing fumes of burning vegetable matter – we cannot trace that back confidently to early prehistory.

A central hearth generates social interaction, but of course its creation also involves cooperation. The accumulation of fuel may be as much of a collaborative activity as hunting, fishing or collecting wild foods. The individual who brings in the most firewood may gain social standing as a result of such achievements.[37]

A regular controlled fire performs other functions that benefit a small group living in close proximity. It is a place to dispose of rubbish, and therefore it contributes to health, as food waste and other discard are burnt – even the disposal of infant faeces, as found at the late Palaeolithic site of Wadi Kubbaniya in Egypt.[38]

Thinking more broadly of the social implications of fire, we should also note the use of fire for punishment. The readiest form of punishment of a miscreant within a social group is exile – removal from the community (in an urbanised community, prison; in other settings, permanent distancing from the group). But other forms of punishment range from the non-fatal spearing in the calf or thigh of Australian Aboriginal tradition to the stigmatisation by a brand from burning or the threat of burning to secure adherence to social norms.

Cooking

The transformation of diet from raw meat, fish and plants to one regularly including cooked food was a major revolution in itself, with multiple implications. While earlier hominins used the opportunity of fire to alter raw foods, this was not the basis of their diet. Once a fire was always available for cooking, human biology as well as economy and society were affected. The change from a diet of predominantly raw food to a diet of cooked food provided numerous benefits.

Cooking kills organisms in food and so makes it safer to consume; important for all but especially important for those most at risk – the sick, the elderly and the young who are being weaned. The availability of safe cooked food can affect the timing of weaning.[39] Given the challenges of successful animal hunting in most environments, Palaeolithic communities were certainly also scavengers, and meat recovered from a predator's kill (or a natural death) would become safer to eat with cooking.

Food preparation might involve placing a food item directly on a blazing fire, or on the embers and charcoal in the hearth once the fire had subsided, or on heated stones. In time there developed the technique of boiling food items in a container over a fire or on hot stones. Earth ovens involve placing the heat above rather than below the item to be cooked. Traditional Aboriginal Australian practices used open hearths, fire-heated stones or earth ovens.[40] The use of fire in cooking may show up in different ways, or be invisible, in an archaeological site.

Cooking can, of course, make edible for humans those plant resources that would otherwise be completely inedible because of their toxicity or because they were too tough for the limits of human dentition. In Australian Aboriginal examples, potentially toxic plants were treated; the kernel of the cycad palm was soaked and roasted and pounded before being acceptable as a food suitable to store for future use.

If cooked food is easier to eat than raw food, it may take less time to chew. In many contexts, this may not be important, but for a mobile group of hunters, cooking food while on hunting expeditions has value. But the same applies for plant foods. Among the Hadza foragers, women travelling to find edible roots would then make an open fire on their journey and lightly roast some to consume before returning with the balance for those in their home camp to eat. Similarly, men would cook and consume part of the results of their hunt on the journey.[41]

Raw meat in warm climates will not last long; cooked meat may last longer; but dried meat can provide a mobile and reliable source of protein as a pre-modern equivalent to refrigeration. The pemmican of Native North Americans, the jerky from Indigenous South Americans, the biltong from Southern Africa and equivalents elsewhere demonstrate the utility of dried meat, and drying over slow fires may be a substitute for the more common sun drying. With meat favoured over plant foods but available less frequently, sun-drying meat was important to the Hadza foragers. Smoking fish to preserve them for future consumption requires the more ambitious construction of a smokehouse.

Thus, cooking is significant for mobile foragers in increasing the range of potential foods: plants that were previously inedible become food, areas that offered limited resources become more productive and the need to move on to new areas for food collecting may be reduced. Although like most foragers the San (Bushmen) of the Kalahari were nomadic, they were recorded as moving their base camp just five or six times a year, finding adequate plant and animal foods and fuel for fire within a radius of six miles (10 km) from the camp (plant foods by weight supplying 60–80 per cent of

The Transformative Power of Fire

food). The territorial adaptation achieved with fire can be compared to the territorial adaptation achieved through biological adaptation in evolution.

The great French mid 20th-century anthropologist Claude Lévi-Strauss identified and emphasised cooking as a significant symbol. In his influential if enigmatic 1964 book translated into English as *The Raw and the Cooked*, he suggested major contrasts and symbolic relations between the activities of roasting, boiling, steaming, grilling and frying, distinguishing levels of culture and nature in each of these processes.[42] Transforming foodstuffs by heat rather than eating them raw is certainly one distinctive mark that differentiates the human state from the animal state.

Biology

Cooking food has a range of different impacts of which increased energy contribution is a primary one: most studies indicate that cooking increases nutrients and improves the calorific return of a foodstuff. This is both from the reduction of energy required to chew and digest the food, and from the contribution in the transformation by digestion. A 30 per cent increase in calorie contribution from cooking has been suggested.[43] However, it should also be remembered that in fatty meats cooking can remove high-calorie fats. And not all 'cooking' is a lengthy culinary process: sometimes it may be just a quick application of fire.

Food provides the body with energy, but some of that energy is used to digest food. Cooked foods can be not only easier to eat but since cooking can make them more tender, they also become easier to digest than raw food. Of the calories produced by protein foods, 20–30 per cent go into digesting it, but more if it is uncooked. It is estimated that digesting meat takes five times as much energy as digesting an equivalent amount of plant-based foods. Cooking increases the digestibility of starch foods significantly, and even brief roasting of tubers increases their ease of digestion and contribution of glucose.

If less energy is required to digest the food, more energy is available for other purposes, including notably the brain. It is estimated that while the human brain represents just 2 per cent of body weight, it uses 20 per cent of the energy of the human body. Here, then, is the marked difference of humans even from other primates and from early hominin ancestors. The body of our species with a relatively small gut and a relatively large brain, combined with the advantages provided by cooking and tenderising our food, means that energy which other primates use in digestion is diverted to serve the functions of the human brain.[44]

This observation has been used to argue for the correlation of human biological development with fire use. The 'cooking hypothesis' discussed earlier in this chapter argued that the biology of *Homo erectus* implied they were already fire users: when cooked food was basic to the diet, hominins required only a smaller gut for digestion and a reduced jaw for chewing, with more energy directed to the brain. While the archaeology does not seem to support this for the Lower Palaeolithic hominins, these features were established with the emergence of anatomically modern humans in Africa. A contrast between *Homo sapiens* and the Neanderthals indicates the latter had a larger and more robust jaw with molars well adapted for heavy chewing. The bigger physical difference is in overall body bulk, considered an explanation for the Neanderthals' highly successful survival through repeated cold periods of Ice Age Europe. When Upper Palaeolithic modern humans arrived in Europe, they could not match the Neanderthals in physical adaptation, and the advantage of their blade stone-tool technology over the prepared-core flaking of the Middle Palaeolithic was not, in itself, sufficient to account for the rapid population displacements. It was the cultural attributes of *Homo sapiens* that gave them advantages, and among these was the ability to create and maintain fire as and when required.

In a competitive situation, this would significantly benefit a group that cooked their food over a biologically similar group that did not. Once fire had been tamed, those who commanded it would gain a strength not possessed by others, giving them an advantage in hunting, food gathering and dominating the landscape.

A domestic fire at the centre of a forager community camp may have negative effects: a smoky environment, especially for a community camping long term in a small, enclosed rock shelter or cave entrance, will have health impacts. Certain woods can harm the foods they cook. It takes time and experience to discover the toxicity from burning woods such as oleander, mangrove, sassafras, oleander, yew, tambootie and laburnum.

Smoke can also have practical and health benefits. If we envisage prehistoric foragers sleeping on the ground using straw or other vegetable matter as a base and perhaps animal skins as a cover, smoking of these removes insects and other pests. Smoking or slow burning of bedding straw was reported from excavations at the Middle Stone Age site of Sibudu Cave in South Africa. At Border Cave where South Africa meets Swaziland, grass bedding was placed on layers of ash from hearths at a date of ca. 200,000 years ago, interpreted as a means to keep insects away.[45]

Food Acquisition

A diet of cooked food may thus provide major advantages, but before food can be cooked it must be acquired. Wildfires created opportunities for hominin forager groups well before they had domesticated fire for their own purposes. Foragers' lives are built around their detailed knowledge of the terrain they occupy and the seasonal cycle of themselves, their food resources and the landscape. Wildfire can move rapidly through grass cover; it kills animals in its path, flushes others out and exposes burrows.

To be successful in adaptation to its environment, any human forager group needs to use whatever tools it can to manage its potential resources. Fire could play a role in landscape management for hunters, just as fire would much later contribute the basis of slash-and-burn agriculture. However, it is important to note that not all deliberate landscape management by fire is directed to food acquisition.

A study of the African Middle Stone Age in Malawi, at a period of more than 70,000 years ago, suggested the use of fire lit in the landscape was a regular part of social strategies. Fire prevented the regrowth of forests, creating a bushland environment with the possibility that this improved hunting resources.[46]

By selective lighting of fires, a group of hunters can improve their access to meat. With careful positioning and starting of a fire, they can force animal prey into a narrow area (natural or constructed) to be speared or even drive them over a cliff. The image of a successful game drive is seen in a number of kill sites in prehistory, where the driving force might be fires lit by hunters, or the noise created by hunters or the use of domesticated dogs.

Smoking ground-dwelling animals out of their underground lairs is an invaluable use of fire: either driving animals from their burrow with a fire at its mouth or suffocating the animal within the burrow before digging it out. Fire is also invaluable in obtaining honey: smoking out bees to acquire a honeycomb to take back to camp. Rock paintings show this method of acquiring honey, such as that from ca. 8,000 years ago from the Araña Caves in eastern Spain, and in the Later Stone Age of southern Africa (Figure 4).

Deliberate seasonal lighting of fires is a feature of many different societies.[47] Fire across appropriate areas was lit and managed to stimulate the growth of new shoots which would attract game. Hunters could then plan their return to achieve a successful kill. In the Kalahari, the bush fires lit by San foragers were a regular feature in the late winter months of August and September.

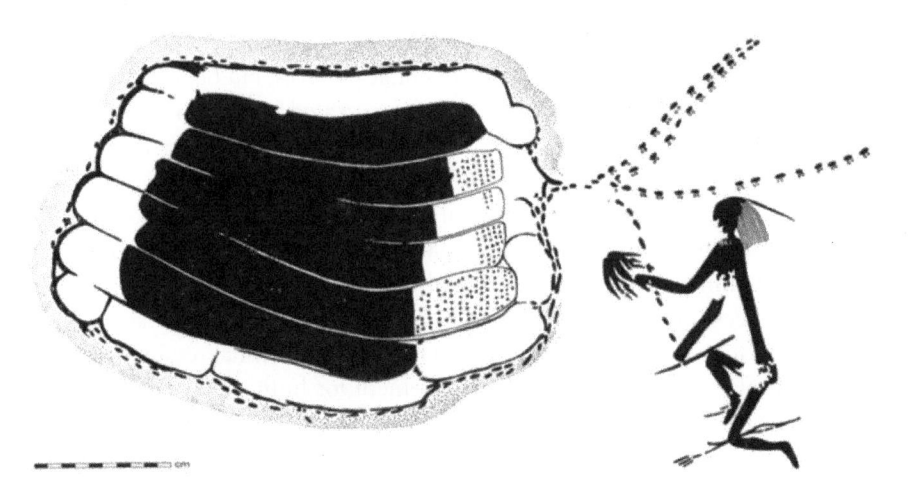

FIGURE 4 Rock painting of a figure smoking a beehive, Toghwana Dam, Zimbabwe (H. Pager, 'Rock paintings in southern Africa showing bees and honey hunting', *Bee World* 54 (1973); 61–68, fig. 6, reprinted by permission of the publisher (Taylor & Francis Ltd, http://www.tandfonline.com).

Fire management of landscapes by traditional Australian Aboriginal communities is sometimes referred to as 'firestick farming': regularly setting fire to grasslands often described as part of a strategy to maximise acquisition of meat from hunted animals. The environmental history of Australia has been linked to this image of firestick farming, changing the landscape over time (Figure 5). The extinction of large fauna in the late Pleistocene has been related by some – controversially and with increasing challenge – to the arrival of fire-wielding humans; while the destructive power of bushfires in modern Australia has been blamed in part on the cessation of traditional Indigenous fire-management practices that followed colonial occupation.

It seems likely that managed fires were often small-scale. Among the Martu of the Australian Western Desert, women burned small areas of 1–10 hectares (2–5 acres) in the winter to expose burrows which may have contained monitor lizards. There were other burns undertaken by hunters, to encourage new growth and thereby attract animals, but these were strictly within territorially defined areas. More broadly, Aboriginal firestick farming reduced the risk of major uncontrollable fires in areas which were essential to the forager economy of the Indigenous occupants: 'cleaning up country', though the relationship of human occupation and landscape burning in the prehistory of Australia is variable.[48] The incidence of greater

The Transformative Power of Fire

FIGURE 5 'Aborigines using fire to hunt kangaroos', Joseph Lycett, ca. 1817 (National Library of Australia, nla.obj-138501179).

fire in the Lake George region from about 130,000 years ago – well before the earliest likelihood of human presence – indicates how complex is the reconstruction of relationships between human settlement, fire and the landscape.[49]

Examples in Australia and in Central America indicate the impact of vegetational change accompanying the first arrival of humans and their assumed use of fire to manage the land and its resources. A study of the California Channel Islands in the USA suggested a marked fire regime affected the region at the time of arrival of the first humans at 14,000 years ago.[50]

The economic impacts from burning the landscape can vary widely by region and location.[51] In the tundra, a fire can destroy the lichens on which reindeer (or caribou) graze, impacting humans who rely on the meat of these animals for their subsistence.

Fire introduced in the landscape can represent a very wide range of functions and intentions. Aboriginal people in the Wik area of Northern Australia during the 1970s gave many reasons for burning, including securing small animals and reptiles for food, clearing the area of snakes but also freeing it from dangerous spirits and opening areas for visibility and access when moving camp. Fire in order to encourage new growth and new grazers was for them not a specific goal.[52] Data collected from a remarkably large set of 231 records of traditional societies (in South and North America and Australia but also Africa, Oceania, Asia, Central America and Europe) presented multiple reasons given for creating a fire away from a home base.[53] The explanations were diverse and bear careful consideration. They were classified as: 3 by accident, 8 chasing away (e.g. flies or bees), 23 clearing the land, 29 communicating, 16 'cultural', 67 game drive, 14 finding foodstuffs, 5 fire prevention, 9 for fun, 1 for illumination, 16 to aid in killing game, 11 to stimulate flora, 21 to stimulate grazing for animals, 8 for warfare. Thus only half of the 231 were directly for hunting animals for food, although of course a fire could have secondary purposes beyond those identified in this analysis. But these activities would rarely be identified in the archaeological record: only, perhaps, where changes in the vegetation of a region (as shown by pollen analysis) changed significantly in a way that regular deliberate burning would cause.

Technology

Before electricity, much technology of the new industrial era was based upon the power of heat (replacing wind, water and traction), and in the metal ages of prehistory smelting and forging were central to the tools of daily life. But the value of fire to ancient technologies extends back into the world of Palaeolithic hunter-gatherers, and even earlier than humanly created fire, when wildfire would be the source.

The basic technology surviving from the Stone Age – flaked stone tools – shows plentiful evidence of fire in numerous sites. Heat treatment of rocks can be identified in archaeological collections by changes of colour, and the context and frequency of artefacts showing the effects of heat will indicate whether this is deliberate treatment by the knappers rather than the fortuitous burning of stone materials. Rocks accidentally fractured by heat can also indicate the presence of hearths.

Heat treatment applied to materials such as silcrete can make them a more suitable raw material for making stone tools. Experiments have shown that heating silcrete nodules before flaking makes it easier to detach flakes and produce thinner flakes and tools with sharper tool edges. To be specific:

> The poorly ordered, strongly interlocking cryptocrystalline fabric of the unheated samples becomes more equigranular and better crystallized with thermal treatment. As a consequence, fractures propagate more readily in heated samples, accounting for their better flaking properties.[54]

The Transformative Power of Fire

Such uses are found in South African rock shelters, as far back as 164,000 years ago at Pinnacle Point. Other southern African sites of the Middle Stone Age confirm that such use to provide raw materials was a normal pattern at some sites from 72,000 years ago.[55] This means that the application of heat treatment to aid in the manufacture of stone tools preceded the sequence of modern human migration from the African continent.

In 2020, Israeli excavators announced they had discovered evidence for deliberate heat treatment of flint to ease artefact manufacture, from the Lower Palaeolithic of Qesem Cave dating before 300,000 years ago. This would make it the earliest site yet to show this process, though it was not necessarily more than an occasional event.[56]

Such processes continued among later prehistoric foragers. Experiments in Australia on both silcrete and quartzite showed the positive value of heat treatment before flaking, which correlated with finds in excavated sites including that at Burrill Lake dating back 20,000 years. Indications are that Australian heat treatment of silcrete dates as early as 42,000 years ago at Willandra but decreases over time, thus suggesting it was a core part of the technology of the earliest humans settling in the continent.[57]

Heat treatment is less relevant where materials such as flint or obsidian were used for stone tools, as these materials can produce high-quality tools with the most detailed retouch and sharpest edges in the hands of a skilled worker.

The use of fire to harden wood is also well attested as part of the process of preparing tools and weapons. Charring the end of a wooden spear can make it easier to sharpen – although, ironically, it might also make it more subject to snapping when in use.[58] Fire hardening of arrows and spears has been widely attested in forager societies. Where bamboo is available, it can readily be hardened in a fire because of its resistance to burning.

Fire was essential to heat and soften the adhesives used to fasten stone spear and arrow heads and other tool heads to wooden shafts. Palaeolithic examples include the Middle Stone Age of Sibudu in South Africa, involving ochre additives in the adhesive gums used from 70,000 years ago onwards.[59] Care was required not just in the preparation of materials but in using just the right temperatures during the process.

A further application of fire is in the smoking of animal hides to prepare them for use in clothing or as bedding by drying, though we cannot identify direct evidence for such a process from early prehistory.

Ritual and Art

Human life is not only about the practicalities of food, drink, shelter and reproduction. Theories abound to suggest how, when and where *Homo sapiens* developed ideas and practices that could be called religious or spiritual or at least non-material, and the use of decorative items among early humans is regularly described as 'symbolic behaviour'.

Often linked to this is the presence of ochre in an excavated site: ochre can be incorporated as a filler with fire-heated resin and gum adhesive used for hafting. Ochre suits decoration, whether of the person, an object or a natural surface. While ochre mixed with suitable fats of vegetable gums serves as a colouring agent on its own, the use of heat can turn yellow ochre to red and can make red ochre darker. So for body painting or other decorative functions, the palette of the painter was broadened by the use of heat.

Prehistoric rock paintings are found throughout the world but among the most dramatic, because of their position as much as their age, are the paintings within dark caves which were created in the Upper Palaeolithic of Ice Age Western Europe. While some of these cave paintings are in areas with an element of natural light, many are in dark spaces where the painters did not live and which their economic activities did not require them to enter. The ability to paint within a cave required fire: a lamp to enter the area of a cave without natural light and a lamp which remained lit long enough for the painters to undertake tasks which frequently demonstrate very great care and allocation of time (Figure 6).[60] Debates continue on whether such paintings in inaccessible locations were associated with shamanic visions, hunting magic, initiation or other purposes. But without reliable lamps, the famous art of the cave sites would not have existed. Examination of the topography of three French caves showed how deliberately light needed to be placed, and a test of the cave site of Ardales in southern Spain showed how a series of carefully positioned lamps would be required to allow the Palaeolithic artists to undertake their work there.[61]

Another experiment created torches from juniper wood, birch bark, ivy, pine resin and deer bone marrow and found these easy to transport and suitable to transmit light in multiple directions while remaining lit for twenty to sixty minutes. They had the disadvantage in a closed space of creating much smoke. By contrast, grease lamps lasted longer but could only light a small area at a time.[62]

One interesting, if rather extreme, hypothesis is that painters who took their torches into deep caves to view their paintings were using fire to

The Transformative Power of Fire

FIGURE 6 Palaeolithic artists in Font-de-Gaume, Charles R. Knight, 1920 (Public domain, via Wikimedia Commons).

deplete oxygen in the enclosed space and by this means induce a transformed and euphoric state.[63]

Lamps were therefore needed to light up the creation of art, but it is not unreasonable to assume that other activities (what we might subsume under the term 'ritual') were associated with the creation of paintings. Whether symbolic non-material activity took place in the darkness of a cave or in the night of an open-air campsite, light and therefore fire was essential.

A famous feature of the European Upper Palaeolithic is the modelled forms of humans and animals – so-called mobiliary art – which are proclaimed as the beginning of sculpture. Some are in soft stone while others are in clay which has survived because of firing: hardened by the action of fire. Finds at the Gravettian site of Dolní Věstonice in the Czech Republic, dated around 26,000 years ago, include sculptures of humans and animals, numerous fragments of burnt clay and structures described as 'kilns'. One analysis has even suggested that the firing was not intended to harden these items to make them permanent but was designed so that they should explode with the heat as a form of ritual. The earliest sculpture known from China, a tiny bird dated about 13,500 years ago, was made on a bone which appears to have been subject to a controlled burning before being carved.[64]

Even earlier is a deer bone from Einhornhöhle in Germany incised with apparently deliberate decoration. It had been softened by fire to make it easier to incise. Given the suggested date of 51,000 years ago and the associated cultural material, this falls within the range of Neanderthal settlement, an intriguing concept.[65]

Low relief engravings on flat stone surfaces ('plaquettes') are a distinct form of European Upper Palaeolithic art. A group of limestone plaquettes from the French site of Montastruc showed evidence of heating and inspired the suggestion that they had been deliberately held near fires so that the shadows cast by low light could reveal the patterns of engraving on them.[66]

There are special rituals associated with death, and careful disposal of the bodies of the dead is a feature of *Homo sapiens* through time and place. Cremation – the burning of the body either with or without the burial of the resultant ashes – developed as a widespread phenomenon but was rare in earlier prehistory. The earliest human remains in Australia, the Mungo I cremated burial of a young adult woman, has been dated at around 40,000 years old.[67] But cremation was not the practice in the European Upper Palaeolithic. In North America, the Marmes rock shelter in Washington State has human cremations dated around 10,000 years ago.

Reverence for fire was common to the values of Indo-European groups who moved into Central Asia, India and Iran in the 2nd millennium BC. Fire was a feature of the Vedic texts that were central to early Hinduism and the cults of Ahura Mazda that would develop into Zoroastrianism. But we can look at many other societies to note the position of fire in belief systems. As with the Indigenous people of the Andaman Islands, it was specifically fire that kept away the spirits of disease and death.[68]

Ideas about early humans might suggest their awe for fire extended to a religious symbolism – something seen in some religious movements of recent millennia – but those ideas are untestable and outside the realm of scientific investigation of the deep past. Fire was part of everyday life, but also part of non-material 'ritual' and what we call mythology. As a scholarly writer on fire has noted about most societies until recently:

> The first act of a day was to kindle a fire; the last act, to bank the coals; and in between, fire was a constant companion. Humanity's power was ultimately a fire power. Anything that so shaped their quotidian world would surely enter into people's understanding of that world and be abstracted into the world beyond.[69]

The Transformative Power of Fire

We can think of some of this when we consider attitudes to fire in modern foragers. The Mbuti of the Congo equatorial forest would light a 'hunting fire ... to rejoice the forest' by the trail they intended to follow on a hunting trip. Or they would build a fire in the hunting camp positioned to indicate the direction of the hunt. When seeking honey, women may scatter sparks from embers on the men as if stinging them like bees. Fire plays an important symbolic role in the *molimo* festival where coals of fire – 'the gift of the forest' – are moved and passed round. The hearth at the centre of a dwelling area is also referred to as a vagina.[70] Fire becomes a means of communicating with a non-material world.

Myths of the Origins of Fire

In 1930, the folklore specialist and pioneering anthropologist Sir James Frazer published his classic book *Myths of the Origins of Fire*, which recorded traditions from societies around the world. It serves as a reminder of the fascination humanity has with fire; the recognition of its importance in all societies; and the role its origins played in imagined histories and in their legendary, divine and supernatural worlds. The many different societies across the world gave widely different accounts of how they acquired fire. This could be from supernatural figures or semi-human ancestors. In the majority of traditions from different continents, fire came to humans through the agency of an animal or bird or reptile. The important thing is that all these societies recognised the importance of fire in their lives; the myths indicate an understanding that it had not always been present and was not a natural part of life but at some point had been acquired, to light them, to warm them, to provide them with the means to have food cooked. Without the gift of fire, the day was short, the night was cold and the food was raw. 'As they had no fire, they could not dance by night and were obliged to dance by day.'[71] Frazer suggested that the pattern of myths implied that people commonly recognised (or imagined) three stages in development: the absence of fire; the ability to use fire but not create it; and the final mastery of how to begin a fire by friction or other means.

These widespread myths of fire contrast with the Vedic Indian references to the sacredness of fire and the Classical Greek story of Prometheus stealing fire from the gods. In the Old Testament, fire can accompany the divine presence of Yahweh, as in the burning bush encountered by Moses, but its possession by humankind is not explained in the creation myths of Genesis.

Conclusion

Once we think of the numerous and invaluable impacts on forager groups from the ability to create fire, we cannot readily accept the idea that past communities voluntarily dipped in and out of the use of fire as an optional resource. When natural fire was available – typically through the occasional and unpredictable occurrence of a lightning-struck fire in the proximity of a group camp – then hominin ancestors before *Homo sapiens* seized the opportunity to capture the benefit. They could scavenge on the meat of burnt animals and hunt fleeing animals or those in exposed burrows or grazers attracted by fresh growth; and sometimes they could take smouldering materials back to provide temporary heat and light and cooking in their own camps. But that is very different from the *taming* of fire: the ability to create a fire whenever and wherever needed and once available to use it for any of the purposes discussed in this chapter.

It does seem that the control and management of fire was a feature of *Homo sapiens* from our species' emergence in Africa through their spread into Asia, Australia, Europe and Asia. Whether Neanderthals had separately acquired the knowledge to create fire remains a matter of debate, but its occurrence in the European Middle Palaeolithic more frequently in hot periods (when wildfires would be commonest) rather than cool periods (when artificial heat would be most advantageous) suggests they probably belong among those who exploited wildfire rather than created their own.

Once the ability to create fire was learned, its advantages would ensure its rapid transmission, and the major advantage of its possessors over others, in competition for resources and territory. A community that can create fire at the time and place of their choosing has gained itself numerous benefits, sufficient to support Darwin's suggestion that the taming of fire provided humanity with the most adaptive advantage after language. Light and heat came under social control and took human life beyond the limitations set by biology and environment. Foragers could transform their foodstuffs and indirectly fuel the capacity of the human brain; and use fire in ever-changing technologies. Art and ritual could develop in new ways. The fire at the centre of the home base provided security as much as a social setting for the interactions of family and group. The taming of fire, with the numerous benefits it brought to human life, marks perhaps the greatest transition in the development of human material culture and society.

Conclusion

If fire transformed prehistoric hunter-gatherer life, it remained at the core of agricultural societies and economies. We cannot imagine the first cities without the light of fires or of cooking processes. The industrial revolution is marked by the use of steam power, where fires converted water into power. Only a century separated the development of gas lighting (lit by fire) to replace candles and lantern from the development of electrical lights, which finally changed the role of fire in human lives.

CHAPTER 3

Domesticating Horses

> No animal is more noble than the horse, since it is by horses that princes, magnates and knights are separated from lesser people and because a lord cannot fittingly be seen among private citizens except through the mediation of a horse.
>
> Jordanus Ruffus, De medicina equorum (The Care of Horses), 1250

Histories of the Horse

Control of fire had brought humans greater control of the environments in which they lived: they were no longer limited by the light of the day, the warmth of the seasons, the edibility of raw foodstuffs, and they were provided with a tool which enabled many social changes. Tens of millennia later, the transition of the horse over time from a source of hunted meat to a domesticated provider of human transport (carrying a rider or hauling a wheeled vehicle) gave humankind further control over their world. The movement of hunter or herder, explorer or trader, soldier or messenger was set at a quite different scale. Social status, wealth and power were signified by control of a horse or a wheeled vehicle well into the 19th century. The taming of horses, the invention of wheeled vehicles, then the linking of the two, had impacts across every aspect of human life, for the individual, the community and broader political power.

Without the horse, human movement was limited by the speed of walking or the use of watercraft. This defined the distance people could travel by foot or alongside a cart hauled by oxen or donkey. It set the rate by which items could be traded overland and the rate by which ideas could be transmitted. It limited fighting strength to what weapons people could themselves carry into battle. All this changed when humans finally established control of the horse, breeding, training and using this unique species, and later using its strength to haul wheeled vehicles. Wild herds of horses had long been hunted for their meat in the Old World; once humans had tamed the horse to ride, they could travel much faster and further to hunt and find new territory, and later to herd other domestic animals. The wheeled cart or wagon, whether hauled by ox, donkey or horse, changed agricultural life and settlement. Horses trained to pull a chariot gave its driver a status in peace and war no weapon could match; and in time, the cavalry horse ridden into battle transformed warfare and power. In widespread societies, an individual's ownership and use of a horse marked social rank and daily convenience for millennia, while the development of the horse-drawn carriage provided transport for others. Those with horses were responsible for major transformations in technology, religion and economy. World history changed as armed horsemen from the Eurasian steppe attacked settled civilisations to the west, south and east, and the horse was a powerful tool in Europe's conquest of the Americas.

The taming of the horse was different from controlling fire. Before the knowledge of how to create fire had been acquired, its applications were known and available when captured from wildfires caused by lightning strikes. The relationship of humans and horses evolved in complex stages: wild horses as a source of meat; tamed horses which could provide mare's milk as well as meat, and which could be ridden to help in hunting down their wild peers and then in herding; the fully domesticated horse, which could be harnessed to a chariot, and later to haul goods in a wagon or cart or people in a carriage; and the horse ridden to travel and trade, for war, sport and status.

The horse was first tamed not in the lands where other wild animals and plants had been domesticated, but in Central Asia in the Copper Age of the mid 4th millennium BC. At the same time, ox-drawn wheeled vehicles were developing further west. It would be as much as 1,500 years later, with the development of the chariot, that the wheel and the domestic horse were brought together; and later still that routine widespread horse riding was core to Old World societies. The control of the horse would dramatically affect the histories of Asia, the Middle East, Europe and the Americas.

The domestication of horses was different from that of other animals. When dogs became part of the human group, they had one economic use: to aid in hunting. Cattle, sheep and goats (like pigs) had wild progenitors which provided meat. Once a community had tamed them, humans could increase their fertility, with selective management of herds, keeping fertile females for milk and breeding and slaughtering males for their meat and skins.

Our image of the horse in human society gives status to the individual rider on their horse, able to travel where they chose on roads or lanes or across cultivated farms or wilder territory. When we read much literature written before the mid-19th century, the horse is firmly in the picture. Characters moved around town and country in horse-drawn carriages, while their goods were carried in vehicles often hauled by horses. The same is true of our images from the medieval or the ancient worlds of Eurasia, when movement, travel and daily life outside of the cities were dominated by the horse. In military narratives, the horse became the most powerful weapon, bringing officers to the battlefield, providing a mount for swordsmen – and earlier for archers – or hauling a war chariot.

Only with the arrival of the steam locomotive in the middle decades of the 19th century – significantly, first nicknamed the 'iron horse' – did an alternative means of land travel begin; the bicycle then the motor vehicle would mean that after the end of the 19th century the horse had begun to be replaced in many regions and in many people's lives.

The power and speed of horses became the measure of travel, of transport of people and of goods, for millennia from the Bronze Age of the Old World into modern times. We hold historical images of later mobile armies of fierce mounted horsemen of the Eurasian steppe descending on the settled lands of farmers and city dwellers to slaughter and conquer, such as the Mongols sweeping through lands extending from eastern Europe to deep within China in the 13th century. The horse gave the invaders from the Old World advantages of power that changed the Americas, while its adoption in the North American Great Plains gave Native Americans a new resource in hunting and fighting.

There have been active debates and often passionate disagreements about the times, places, processes and uses of the first horse domestication.[1] The archaeological evidence from prehistoric societies is largely indirect, and open to different interpretations, but locating the first domestication of wild horses within the steppes of Eurasia and in the 4th millennium BC remains most widely accepted.

Broadly contemporary with the taming of horses, wheeled vehicles were being developed elsewhere to complement and replace the limits of transport by boat, sled or human strength. Here too the archaeological evidence leaves ambiguities, with carts or wagons appearing in Europe, the western steppe and the Middle East in the mid-4th millennium BC. With the crucial development of the wheel to equip a cart or wagon, it was not horses but other animals that were used first for traction: oxen and donkeys. After 2000 BC, the horse was first brought together with the wheeled vehicle in the form of the two-wheel chariot, hauling these lighter-weight vehicles in both the steppe and the early civilisations of western Asia to the south. Gradually, the speed of the horse was combined with its strength to haul both goods and people in wagons and carriages, but the power and speed of a horse with a single rider would become a long-term trend in civilian and military life.

Horses before Domestication

The Indigenous societies of the Americas would be changed forever after the 15th century by the arrival of the Spanish and others with their horses and horsemanship. Empires would fall in South America and Mesoamerica, and land would be lost to the invaders and new settlers. Trade in horses to Native Americans would alter their inter-group relationships and their economy.

Yet these horses introduced from the Old World had a deeper ancestry in the New World, where they were no longer found in the wild. The ancestors of the horse family had moved from the New World to Siberia during the Pleistocene, crossing what is now the Bering Strait on land exposed at periods of low sea levels. Some movement may have continued in both directions until at least 26–20,000 years ago.[2] Wild horses continued to exist in Alaska, genetically related to those from Siberia, until they disappeared around the end of the Pleistocene some 12,000 years ago. Their extinction may have been influenced in part by human hunting: bones of butchered horses are a feature of a number of camp sites of the first forager cultures of North America.[3]

In the Old World, the ancestral horse diverged into different species, including the zebra and the wild ass (domesticated as the donkey) in Africa, and the onager and the horse in Eurasia. We encounter the latter in the human record long before they were tamed, as one of the hunted prey of our Palaeolithic ancestors found in the food-waste bones at archaeological excavations, as well as shown in prehistoric art.

At a site at Boxgrove in England, *Homo heidelbergensis* in the Lower Palaeolithic of some 480,000 years ago had carefully butchered a single horse whose shoulder blade had an apparent spear wound. Wooden spears at Schoeningen in Germany, associated with either *Homo heidelbergensis* or Neanderthals at a little over 300,000 years ago, appear to have been used to kill wild horses whose bones show their butchery marks.[4]

Horses were a significant food of Europe's Neanderthals in the Middle Palaeolithic. They were an important contribution to the diet of our own species *Homo sapiens* in Europe's Upper Palaeolithic. Foragers of the Ice Age acquired and consumed a broad range of plant foods, and bones from their occupation sites demonstrate the varied range of wild animals eaten, large and small. Yet the contribution to a community's diet from killing and butchering a horse was substantially greater than most alternatives. One horse would provide plentiful meat to a community: 'more than a hundred hares'.[5] In many sites of the European Upper Palaeolithic, the frequency of horse bones compared to those of other large animals suggested they were either commoner prey or easier to catch. Evidence from some Upper Palaeolithic sites in Europe has been interpreted to indicate that after herds of females or of bachelor males were ambushed, horses were butchered at the kill site with some consumption there, and heads valued enough to be taken away to a base camp.[6]

The remarkable Upper Palaeolithic site of Solutré in France has vast quantities of bones from thousands of horses hunted over a period of 20,000 years. These may have been driven over a cliff or more probably chased and trapped near a cliff before killing.[7] Yet the paucity of marks on the bones suggests that butchering for meat was only partial and selective, though we can assume the skins of the horses would have been readily used.

Dramatically, we can see horses depicted in the prehistoric rock art of Europe's Upper Palaeolithic hunter-gatherers (Figure 7).[8] More than 1,250 images of horses (sometimes painted alongside bison or ibex) have been found among the cave paintings of Ice Age western Europe, created between 31,000 and 11,000 years ago. In fact, horses are the most common species depicted, making up 30 per cent of all mammals in painted cave art, often with carefully observed details.[9] This emphasises their importance in the perceptions of the hunters, including images showing aspects of the hunt itself. Some horse paintings, especially later works, are relatively naturalistic, while others have stylised and exaggerated elements.[10] Although the outward form of these Upper Palaeolithic horses has resemblances to the wild Przewalski horse (*Equus ferus przewalskii*) known today in the steppes of Central Asia, these were not ancestral to the horses domesticated in Europe.

These paintings allow us to see the colours of horses' coats. Brown and black are most common in the art of Lascaux and Chauvet, with a rarer dappled colour in horses on the cave walls of Pech-Merle in France, dated to ca. 25,000 years ago. It has been confirmed through DNA studies from bones that this dappled colouring, as well as brown and black, was present in wild horses in Europe before their first domestication; the colouring was not just from the imagination of the artists.[11]

Histories of the Horse

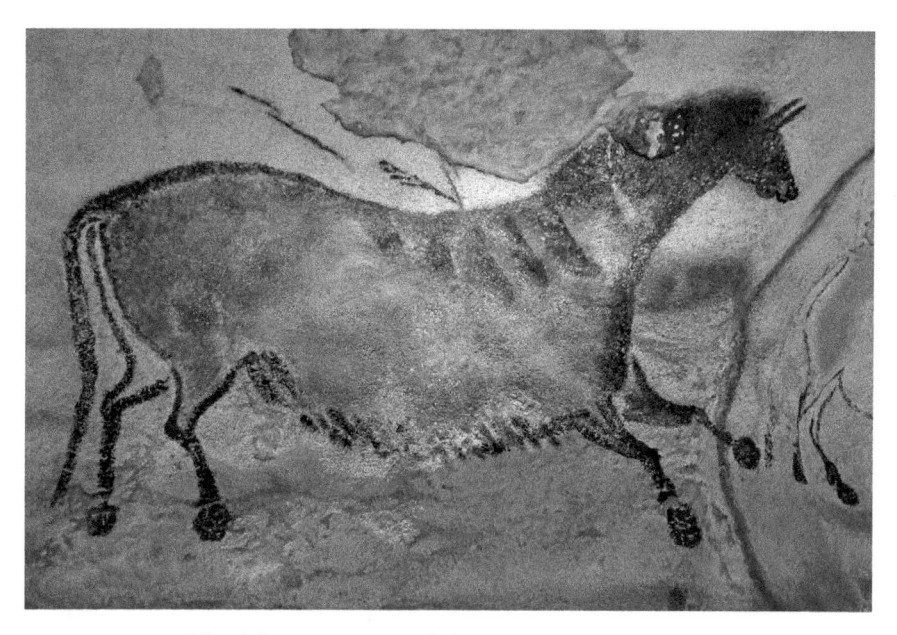

FIGURE 7 Upper Palaeolithic cave painting of a horse at Lascaux (photo credit: Bonnafe Jean-Paul/Moment/Getty Images).

Dating even earlier than the European rock paintings is a 35,000-year-old sculpture of a horse carved from ivory, just 5 cm long, found at Vogelherd Cave in south-west Germany (Figure 8).

But the changing vegetation patterns of Europe after the end of the Ice Age, together with possible over-hunting of the horses in remaining habitats, meant that by the time agriculture reached Europe, wild horses were relatively scarce.[12] Where they did thrive and remain numerous was the grassland steppes of Eurasia.

Horse Biology and Domestication

The prehistory of horse domestication can be examined from the archaeological evidence at excavated sites, but also by comparing the DNA of modern horses with that from the bones in those sites and interpreting how the gene pool of domesticated horses changed. Geneticists have sequenced 300 horse genomes, 'the largest genome-scale time series published for a non-human organism'.[13] Debates inevitably remain about how the genetic evidence fits the pattern of prehistoric cultures.

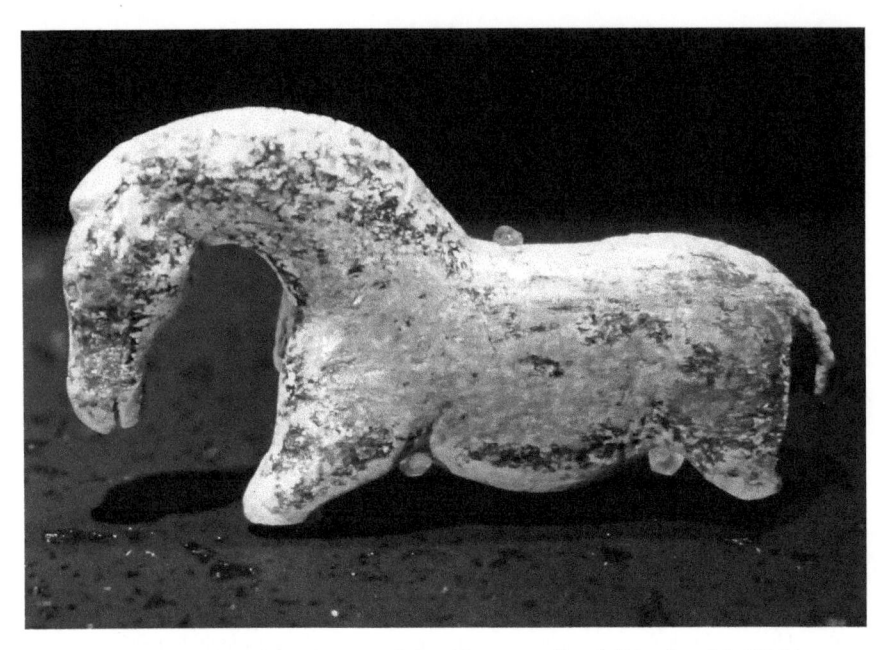

FIGURE 8 Palaeolithic horse sculpture from Vogelherd (Wuselig, CC BY-SA 4.0, via Wikimedia Commons).

The early stages of horse domestication did not necessarily involve trading domesticated stock or imply migration with tamed herds. Awareness of the means and advantages of horse taming could spread – the idea, not the animals – and be applied by new groups of horse hunters to capture and tame their own stock.

The DNA of domestic horses indicated a relatively homogeneous gene pool of wild horses in the Late Pleistocene, the terminal Ice Age of Eurasia.[14] The horse bones which archaeologists recovered from human occupation did not show marked changes in body shape or bone structure which would demonstrate human control of the horse.

With genetic contributions from different regions, it seems that that wild horses continued to breed with domestic stock well beyond and after their first control.[15] Perhaps surprisingly, a wider variety lies in females rather than males; this is not a scenario of tame mares impregnated by wild stallions. The genetic pattern suggested that once there was a continuous breeding population (rather than just taming of individual wild stock), then it was mainly males (stallions) that were maintained selectively from generation to generation, breeding with a much broader range of wild mares.[16]

As the idea of taming horses spread, herds may therefore have been augmented by capturing wild females rather than the more difficult challenge of capturing male stallions. Genetic studies imply that a reduction in the number of stallions used for breeding has continued as a deliberate pattern for the last 2,000 years, as horse stocks have been strengthened with selective breeding, although a survey notes that ancient Roman breeders were 'a lot less choosy'.[17]

Genetic studies relating to horses' coat colours suggested some deliberate breeding for favoured colours. Wild Pleistocene horses appear to be most commonly bay (brown) in colour, and later (but still before domestication) horses in the Iberian peninsula included both bay and black horses. From the 3rd millennium BC, chestnut horses were found in Siberia, and more variety developed in the European Bronze Age, seemingly the result of selective breeding.[18]

Significantly, it seems from genetic evidence that the first horses likely to have been tamed, those of mid 4th-millennium BC Botai in Kazakhstan, may be ancestral to today's wild Przewalski horses of the Central Asian steppe, now thought to be largely descended from once-domesticated stock.[19] A different ancestry of our modern horses apparently lay in the western steppes, and especially in the Volga-Don region, before the 3rd millennium BC.[20] In the spread westwards from ca. 2200 BC, these domesticated horses reflected previously wild populations of that region.[21]

The Horse before the Cart: Botai and the First Taming

Where and when did humans first learn that horses had more to offer than a good meal for all? Although the wild horse provided the Upper Palaeolithic hunters of Europe with a significant source of meat (and art!), we look millennia later and to the steppes of Central Asia for the possible location of the earliest domestic horses, at a date after 3500 BC. The evidence for such tamed horses is circumstantial and open to conflicting interpretations.

The earlier domestication of cattle, pigs, sheep and goats that spread across the Middle East, Europe and northern Africa had begun at the north of the 'fertile crescent' region around 9000 BC, between southern Anatolia in Turkey and western Iran, where the world progenitors of these domestic animals lived. China would domesticate pigs independently about two millennia later. The horse remained untamed in the main areas where this farming began.

While farmers in Europe were developing the ox-drawn cart and wagon in Europe in the middle of the 4th millennium, a different process was quite probably occurring quite independently to the east in the Central Asian steppe: the beginning of horse taming. Here, horses are thought to have been controlled to be ridden as an aid to hunting basic food animals – and those food animals could include, especially, wild horses.

The steppe, 6,400 km (4,000 miles) long, is the open grassland zone found south of the Siberian forests and north of the Central Asian deserts and the inland Aral, Caspian and Black Seas. In much of this exposed landscape, very cold temperatures in winter are matched by high summer temperatures, with occasional extremes. To the east (past the Altai mountains), the steppe extends to Mongolia; and to the west, it reaches into south-eastern Europe, with a further presence in Hungary. The boundaries of the environment encouraged trade and movement on a west–east axis in prehistory, similar to the Silk Road of historical antiquity, and it seems that knowledge of horse husbandry may have first spread along this axis.

The horse was native to the steppe grassland, being able to survive its extremes of temperatures and find grazing on frozen ground. In a period of cooler, drier climate, the Piora Oscillation from the late 5th to mid-4th millennia, the grassland area of limited vegetation had increased, and its resources favoured the horse.

Human groups hunted across the steppe, and horses formed a focus of their movement and economic strategy. Many (though not all) archaeologists consider there to be enough indirect evidence to show that horse hunters of northern Kazakhstan at the heart of the steppe developed control of some of their prey, corralled to ride, breed and to provide milk. These proposed pioneering horse tamers were of the Botai culture from ca. 3500 to 2900/2700 BC, part of the later Copper Age (also called the Eneolithic or Chalcolithic).

Horses domesticated for riding would enable the hunters to pursue their prey directly, beyond the need for trapping or ambushing them. A mounted hunter could travel much further from home in pursuit of prey and carry more of the meat back home, with the slaughter of corralled stock as a supplementary source of meat.

The Botai people were semi-sedentary, with a diet dominated by meat rather than plant foods.[22] They occupied settlements of turf-roofed dwellings partially sunk into the ground. These houses were typically six metres across and often square in shape. As many as 250 houses have been found at individual sites, not all inhabited at the same time. Large permanent villages might appear inconsistent with an economy dependent on hunting mobile herds of horses, but riding of tamed horses would allow them to hunt further from their homes and also to use milk and meat from domestic stock.

They made pottery, but metal was rare, and they relied largely on stone or bone for their tools. Many objects at Botai, including spears, decorated items and stamps for decorating pottery, used horse bones as their raw material.[23] The bone tools included punches and forms suitable for processing leather from horse hides, such as items suitable for horse-control harnessing.[24]

Markings between the cheek teeth and the forward teeth on some of the horse jaws found at Botai have been interpreted as evidence of domestic control. Such marks can result from mouthpieces (leather or rope bits or bridles used to control them either led or ridden) when used over an extended period to manage the movement of a horse.[25] The presence of corrals suitable for holding horses, and horse dung collected and stored in particular locations, supports the image of domestication.[26]

Of the vast quantity of animal bones found at the site of Botai, almost all were from horses. At another Botai-culture site, Krasnyi Yar, 90 per cent of the animal bones were from horses, and most of the rest from aurochs (the extinct wild cattle). The age and sex of the horse bones (with males slightly exceeding females) suggest a mixture of domestic and hunted horses.[27] The skulls and vertebrae indicate whole carcasses were butchered in the settlement, whether from domestic animals or carcasses hauled by riders of tamed ones.[28] Head wounds on some horse skulls imply they were killed by a poleaxe blow to the head, typical of modern slaughter. If horses had only been killed at a distance, carrying back just favoured parts would be more likely.[29]

At Botai-culture sites, horse skulls and bones were buried in some of the many pits found outside houses – sometimes associated with buried dogs.[30] One burial had four humans surrounded by the skulls of fourteen horses, with other horse bones set in an arc.

A mare in captivity producing milk for a foal could be milked for human consumption, as in the modern Central Asian steppe. Fatty deposits on Botai pottery may represent mare's milk – something that could only be obtained from tamed animals.[31]

We can suggest how the first horses were tamed, with orphaned or other captured foals brought into the community.[32] If a foal remained corralled, rather than consumed for meat, its potential to be ridden could be tested; the subsequent development of leather ropes and the horse bit served to control it.

The horses of the Botai culture were not ancestral to those found further west.[33] So, if we accept the arguments that the Botai people had domesticated, milked and ridden some of their horses, it would be the *idea* of horse taming that spread westwards and in due course eastwards along the steppe regions where wild horses were present, with communities separately taming their wild hunted prey. By 2000 BC, it was routine to breed and keep horses to the east of the Urals.[34]

The Copper Age Botai culture of the central steppe, with horses its favoured source of meat, therefore presents a possible context for selective domestication at around 3500 BC. After 2500 BC, the economy of the Kazakhstan steppe moved to cattle and sheep husbandry, with horses used to herd them.

Domesticated Horses in the Western Steppe

It is a persuasive idea that in the 4th millennium BC Botai-culture horse hunters of northern Kazakhstan tamed some of their prey for meat, milk and riding, and that the idea of horse domestication may have spread westwards from there to be adopted by Copper Age societies of the Pontic-Caspian steppe. Lying west of the Ural mountain range, this region extends across the north of the Black and Caspian Seas, including modern Ukraine and parts of southern Russia, to the mouth of the River Danube. It was from this area that raiders on horseback – Scythians in the 7th to 3rd centuries BC, and Huns in the 4th to 6th centuries AD – would much later emerge to attack urban civilisations.

Some scholars have interpreted finds from the mixed agricultural Copper Age communities of this western steppe region as evidence of independent domestication of the horse. If they had domestic cattle, sheep and goats in the 5th millennium BC, could they also have tamed the horse they hunted for their meat? If so, could the idea of taming even have spread *east* to Botai from there? Arguments on the timing of locally domesticated horses use circumstantial evidence from animal bones, material culture and funerary rituals and are subject to continuing debates.[35]

In the early Copper Age of Russia's lower Volga region from the 5th millennium BC, human burials were accompanied by the bones of domestic sheep and cattle and also of horses. The material culture included carved figurines of horses as well as cattle. The symbolism could mark the significance of someone successfully hunting down a wild horse to supplement the normal economy of domestic animals and cereals or reflect the possession of tamed horses.[36] Some finds have been identified as cheek pieces from horse harnessing, although other researchers have doubted such an interpretation. Cheek pieces are the elements that would be used in a horse bit at the corner of a horse's lips to hold a mouthpiece in place, and which put pressure on the lower jaw.[37] Mace-heads often interpreted as representing horse heads are found as early as the 5th millennium BC within the Copper Age of south-east Europe, which shows signs of increasing contact with the steppe.

Important evidence in these discussions comes from the occupation areas and burials at Dereivka (Deriivka), which lies in an area of steppe not far from mixed forest-grassland on the Dnieper River in Ukraine. Belonging to the Copper Age Sredny Stog culture, the settlement dates before 3500 BC and is therefore older than Botai. Animal fats on pottery failed to provide evidence for the consumption of mare's milk. Carved antler items have been suggested as possible cheek pieces.[38]

Hoe agriculture of cereals seems to have complemented the meat from cattle, sheep or goats, pigs, fish and wild animals including deer, but horses provided more meat than any other animal.[39] The horse bones showed a predominance of males, which could imply selective hunting, or the maintenance of a herd of females for milk or breeding. Most horses were killed when aged between five and eight years, which may indicate a mainly wild rather than domestic source.[40] A cautious interpretation would be that Dereivka people were hunting wild horses to supplement a diet of domestic plants and animals. More examination of the bone material for evidence of bit wear on jaws might change this perception.

Further circumstantial evidence on early domestic horses relates to the Bronze Age Yamnaya people who occupied the Pontic-Caspian steppe between about 3200 and 2600 BC. Known for their early wheeled wagons (hauled by oxen, not horses), their graves contained butchered horses yet no buried harness equipment. Animal heads including possible horses are found on 'sceptres' or maces, and images of horses appear but without a rider or a cart. The apparent presence of milk from horses in the diet is evidence to support their domestication.[41] Other dairy products were part of the Yamnaya diet and were associated with their westward spread.

Alongside this indirect evidence for domesticated horses, an interdisciplinary study of human remains from Yamnaya sites has proposed that the skeletal remains of five people from the Yamnaya period show evidence of physical changes of the type that can arise from regular horse riding, with other burials also showing possible indicators of riding.[42] The study asserts that this is the earliest firm evidence for horse riding anywhere in the world, arguing that the earlier horses from Botai may have been domesticated only for their milk. Genetic studies of human remains showed no links of the Yamnaya people with the Botai people. Nor were there genetic links between the Yamnaya of the steppe and the Hittites and other Anatolians who introduced the horse into the Middle East – genetically these showed descent from local foragers. This reconfirms that a spread of the idea and methods of horse herding was not necessarily linked to migration by horse herders themselves.[43]

The analysis of a large sample of horse genomes from skeletal remains found in archaeologically dated sites concluded that the movement of horses from the Volga-Don region of the Eurasian western steppe dates from after 2200 to 2000 BC, later than the western movement of the Yamnaya.[44] The firm conclusion was that there was no connection between the date of a possible Yamnaya expansion westwards and the spread west of the domesticated horses ancestral to subsequent horse populations. These ancestral horses were found to the west, in Anatolia and the Danube basin, and east, in central Asia, after ca. 2200/2000 BC.[45]

While the steppes of Eurasia seem established as the location where wild horses were first tamed and first ridden, there is no consensus on the time and place of this major development in human control of their world. But the same area was also marked by the development of early ox-drawn carts and wagons; and then, in the western steppe around 2000 BC, the wheeled vehicle and the tamed horse would be combined as the chariot.

The Cart before the Horse: Animal Haulage by Ox (or Donkey)

The cart and wagon developed quite independently from the domestication of the horse. Without a wheeled vehicle or pack animal, pastoralists moving with their herds could only take what they could manage to carry (including carrying their smaller children), as could mobile foragers. Settled farmers needed to haul their produce in baskets, or in bundles of grain stalks, or hauled by tumplines – straps fastened on the forehead – as used to transport timber over long distances for the buildings of Chaco Canyon in New Mexico.[46]

Traders were limited by what they could carry to exchange. A sled with ropes would assist in moving goods: sleds are known from South Russia's Copper Age by 4000 BC, and a millennium later in Mesopotamia (drawn by donkey or ox); a sled accompanied a human burial in Dorchester, England from the mid 3rd millennium. Sledges developed even earlier for transport on Arctic snows.[47]

Before the wheeled wagon, watercraft offered an appropriate means to transport people and goods. We know that early anatomically modern humans did manage to cross water – from eastern Africa to the south-west tip of Arabia and across Indonesian islands to New Guinea – though this does not imply a widespread or routine capability to make or use watercraft. In prehistory, simple rafts could cross rivers; dugout canoes could be navigated on inland waters, exploit coastal resources and travel between adjacent mainland and island locations. The bluestones at England's Stonehenge, transported from south Wales around 3000 BC, required human-hauled sleds on land but may have also involved water transport around the coast. The early urban civilisations based on the Nile, Tigris and Euphrates, Indus and elsewhere could travel and trade widely by river and sea.

The invention of a two-wheeled cart and a four-wheeled wagon pulled by domestic animals was so revolutionary an idea and such a benefit that it spread rapidly; indeed, so rapidly that it is not possible to agree on an exact location for its origin. Humanity harnessed already domesticated animals for haulage, the wheel to turn this power into forward motion and the cart to use this forward motion and move goods from place to place.

The wheeled vehicle appeared about 3500 BC and is known from widely differing social and economic contexts, from Mesopotamian cities to small northern European farming villages, in the Caucasus and western steppe, and after a few centuries as far as the Indus. The horse-drawn cart is a familiar concept, and part of our image of the not-too-distant past, but the cart was first developed to be hauled by the domesticated ox and adapted in the Middle East to use the donkey.[48] A single ox (typically a castrated bull) can be harnessed to a cart, although a pair of oxen add strength and aid the process. Oxen are stronger for haulage than horses, but horses are faster. Once horses were used for haulage, two in harness aided control.

One major impact of the wagon or cart was that, since it could carry produce further than people alone could, it allowed farmers to live a greater distance from their fields. Instead of scattered homesteads, more concentrated groups of individuals in small villages could more readily bring their crops home. The positive features of village life developed, as did the defensive advantages of closer fortified concentrated settlement.[49] The wheeled vehicle developed broadly at the same time in European agriculture as did the hand-hauled plough, which in turn allowed a larger area of planting than a hand-held hoe.

A cart required carefully formed wheels that could be fixed to rotate freely but not too loosely on an axle, unless the axle itself was positioned beneath the cart to rotate. Carts required axles fastened to a constructed open container; pole and yoke equipment to fasten the box to the animals for haulage; animals which would follow directions to pull; and at least a rope to control their direction and movement.

The first cart wheels were of solid wood: it would be long before lighter-spoked wheels were developed. The wheel might be a single piece – in north-west Europe, a large old oak or alder tree was used to make a wheel 80 cm (30 in) across. Three shaped wooden pieces fastened together could provide the same resilience.[50]

Steerage was more of a challenge. A two-wheeled cart with wheels turning independently on an axle could make turns if these were not too sharp. A four-wheeled vehicle – that is, a wagon – with two fixed axles was much less manoeuvrable, and changes of direction required complex to-and-fro movements. The first steering axles only emerged in the mid 1st millennium BC, in Germany and Italy.[51] Steerage harnesses then came into their own.

The weight of a wooden cart or wagon with solid wheels could already be significant before it was loaded with produce or goods, though a four-wheeled wagon could carry a heavier load than a two-wheeled cart. A prehistoric wagon from Holland was estimated to weigh 670–700 kg (1470–1550 pounds) empty: good reason for harnessing two oxen rather than one.[52]

The idea of a wheeled cart or wagon was copied from society to society – perhaps as traders moved across broad regions. There was no need for carts themselves to be traded: indeed, no practical possibility given the absence of roads wider than foot tracks penetrating the forests of prehistoric Europe or across the landscapes of the Middle East of 3500 BC. There was no need to trade haulage animals, as the cart idea was adopted by groups which already had domestic cattle (and when the idea was applied in the Middle East, donkeys).

Images of wheeled vehicles appear in the Uruk period of Mesopotamia (Iraq) and in tracks below a burial mound in Flintbek, northern Germany, both dated around 3500–3400 BC.[53] The Uruk images are pictographs in archaic texts which appear to show a wheeled cart. These are commonly interpreted as four-wheeled wagons viewed from the side, but as the symbol means 'two', this could be a two-wheeled cart with both wheels shown. It is not impossible, however, that these were images of sleds moved on rollers.

Vehicles with two wheels and with four wheels in both civil and military roles are well attested from the Early Dynastic Sumerian city states of

FIGURE 9 Wagon from the Royal Standard of Ur (Public domain, via Wikimedia Commons).

Mesopotamia in the 3rd millennium BC, and from western Iran and Syria in the same period.[54] A wagon for military use with one armed person in front and one behind is part of the frieze on the inlaid wooden box often called the Standard of Ur; this was found in an Early Dynastic royal tomb dated ca. 2500 BC (Figure 9). It seems more likely that the axles on these vehicles were fixed rather than turning with the wheels. On a wheel from Susa in western Iran, the wooden construction was pierced externally by copper nails. The two-wheeled cart was controlled by a man either straddling a pole or seated on a bench. The four-wheeled wagons could have higher barriers to hold.

The later centuries of 3rd-millennium BC Mesopotamia adapted these models and approach for both two- and four-wheeled vehicles, with some improvements to harnessing the draught animals for control. Civil rather than military use then appeared the norm.

Parallel deep wheel impressions were found within a long barrow burial mound of the Funnel Beaker period (TRB Neolithic) at Flintbek in Germany; these appear to represent multiple to-and-fro uses along the same track, about 20 m long. The wheels would be about 1.10–1.20 m apart, and just 5–6 cm (2–2.5 in) wide. The relationship of the tracks to the burial mound above it remains an enigma.

At Bronocice in Poland, decoration incised on a pot from the Funnel Beaker period of similar date, around 3500 BC, included symbols thought to represent a four-wheeled cart but with a similar circular symbol in the centre – nicknamed a spare wheel!

Dated not long afterwards are model wagons found in the middle part of the Danube area, such as that from a grave of the Copper Age Baden culture of central Europe (dating from ca. 3300 to 3100 BC onwards) from Budakalász in Hungary. Built on four solid wheels, it showed a connecting shaft rising from the lower part of the square vehicle.

Models and depictions of wheeled vehicles are found north of the Alps between 3500 and 3100 BC. Their rarity in burials suggests they were prestige items, with an almost symbolic respect for the innovation. From 3100, these depictions cease, after which wooden cart or wagon parts can be found, including what may be wheels buried as part of a ritual.[55]

Knowledge of the wheel and ox-hauled cart or wagon also spread eastwards. Early carts appeared after about 3100 BC in the north Caucasus region and the Pontic-Caspian steppe of southern Russia and Ukraine. There are finds from within the Yamnaya culture (ca. 3200–2600 BC) and the subsequent Catacomb culture (ca. 2700–1900 BC), both to the north of the Black Sea.[56]

People of the Yamnaya (Pit Grave) culture used a bronze and copper technology and buried their dead in *kurgans* (mounds over pit graves), and their settlement pattern suggests variable reliance on mobile pastoralism and some crop agriculture. In places, horse meat was significant in the diet. If some horses were tamed, this would be for milk and meat and to ride in herding and hunting, not for haulage.

More than 250 two-wheeled carts and four-wheeled wagons have been identified in excavations from Bronze Age *kurgans* in the western Eurasian steppe, with increasing sophistication over time.[57] Within elite *kurgan* burials were four-wheeled wagons, such as at Kuban, with solid wheels rotating on a fixed axle. Such burials did not usually include the draft animals, the oxen that pulled them. Perhaps it was the vehicle that marked status, while cattle were owned more widely. The steppe wheeled vehicles were typically 2.0–2.5 m long with a distance across the axle between the wheels of 1.45 m (57 in), a gauge not very different from that of Roman carts.[58] Such a size was not suited to carrying very large loads.

The westward spread of the Yamnaya people into Europe has sometimes been linked to the spread of Indo-European languages. Language terms traced back to a Proto-Indo-European because of their wide distribution in derivative languages include terms associated with sheep, cattle, pigs and wagons (wheel, axle) but not chariots.[59]

Further east, from the late 4th to early 3rd millennia BC, models of two-wheeled carts were found at Altyn Tepe in Turkmenistan, Central Asia. By the end of the 3rd millennium, cart design here had developed into four-wheeled wagons.[60]

One interesting innovation for some wagon wheels has been noted from sites in Iraq and Iran as well as Turkmenistan from the later 3rd to early 2nd millennia BC. This was the use of segments made of copper or bronze clamped on to create a form of tyre around a solid wooden wheel, a costly addition which increased the overall wheel weight; this innovation did not persist.[61]

While the ox to haul a cart was the norm in Europe and in the Eurasian steppe, a common pattern for haulage in the ancient Middle East was use of the donkey (*Equus africanus asinus*). Recent genetic studies have suggested the origin of the domesticated donkey lies in north-east Africa, with an estimated date about 5000 BCE, when pastoralists tamed the wild ass to help with herding.[62]

Donkeys were present in Egypt by the 4th millennium BC at the latest, and at the same time Copper Age figures from the Levant show donkeys carrying produce in panniers.[63] Panniers are of course another form of transport, where the donkey has advantages over the ox (or horse). Bone evidence from donkeys at the very beginning of Egyptian civilisation show the impacts of bearing heavy loads for their owners.

Seal impressions from Tell Brak and Tell Beydar in Syria show donkeys (possibly donkeys crossed with wild onagers) hauling four-wheeled wagons by about 2250 BC. At Tell Brak, skulls of donkeys showed marks on jaws from bit wear which implies they were fastened for regular use in haulage. The donkey would long remain a favoured animal in rural labour. Horses and donkeys are mentioned together in Mesopotamian texts by the 21st century BC. A genetic study of the equids called *kunga* used for haulage in the region in the mid 3rd millennium showed that their source lay in deliberate breeding between female domestic donkeys and male wild asses.[64]

Horses and Chariots

While the horse was tamed on the steppe and ridden while herding domestic animals or hunting and to provide meat and milk, the wheeled vehicle to carry goods was developed to the west, hauled by oxen (and in the Middle East by donkeys). What first brought horses and wheels together was the development of a specific innovation: the two-wheeled chariot to carry either one or two men, with its speed achieved by a single or two harnessed horses.[65] Built with spoked wheels, such vehicles were much lighter than the solid wheeled carts for transporting goods that preceded them.

Evidence for early chariots includes the buried vehicles themselves (often with impressions of the spoked wheels visible in the earth), harness and bridle parts among grave goods and the bones of sacrificed horses, as well as visual representations.

'Chariots' is a shorthand term for horse-drawn two-wheeled vehicles built to carry people and to do so at a pace that could not be achieved by oxen or donkeys. Only some chariots would be used in war: their military contribution was not necessarily their primary purpose, and in time some would be used in sport. Early chariots could be used as marks of status and prestige, allowing their owners to travel widely and to do so in the public gaze. It is not surprising that the invention of the chariot spread widely and lasted long.

The construction of a chariot required specialist skills to construct a vehicle light enough to haul with light spoked wheels and an axle fixed to the chariot body in such a way that movement was not impaired. It involved a structure to fasten horses to the vehicle and equipment to control them. The best wheels would be made using bent wood, glue, leather and possibly metal. So the technology of chariots required experienced experts in carpentry, leatherworking and sometimes metalworking.[66]

Use of chariots also required different skills. It needed horses bred in captivity, then trained to haul the vehicle and follow instructions; careful connection of horses to the chariots they pulled; harnessing to control and guide the horse or pair of horses; and not least chariot drivers skilled in controlling horse and chariot, and perhaps fighting. Additionally, the right fodder was required to maintain the strength of the horse – a feature that would be emphasised in texts from the Middle Eastern states that possessed them. Once these were all mastered, the driver of the chariot had major advantages over anyone on foot. If a chariot was used for military purposes, additional skill levels were essential.

The combination of specialised skills involved exchange of goods and services and would privilege the more elite members of Bronze Age society. But were the chariots and horses found in burials used for fighting, for exhibiting prestige, for use in an afterlife or for all these roles?

The handling of a chariot clearly required an effective bit to manoeuvre the horses.[67] Harness and bridle parts found among grave goods (including bone disc-shaped cheek pieces) demonstrate means of control. Cheek pieces are often found with buried horses after 2000 BC. The presence of cheek pieces indicates that other parts of the harnessing were of organic materials (leather and rope) which have not survived.

Since chariots appear during the 2nd millennium BC in western steppe societies, in the southern Urals and in more advanced states of the Middle East, who learned from whom? Domesticated horses and ox-drawn wheeled wagons coexisted in the steppe. Radiocarbon dating would support the possibility that the horse-drawn chariot originated in the Bronze Age communities of the steppe in the early 2nd millennium BC and was then readily adopted by the urban civilisations to the south.[68]

Numerous burials of humans with early chariots have been found from the Bronze Age of the Volga-Don area of the northern steppe and forest steppe and from the steppe south-east of the Urals (such as the Sintashta culture), dating to the early 2nd millennium BC. These burials were in *kurgan* earth mounds with skulls of horses (of biologically modern type), chariot wheels and bridle parts. Such chariots with burials may represent a means to convey the deceased human to another world after death.[69]

The Sintashta communities were reliant on pastoralism of cattle, sheep and goats but also on cultivating crops. Sintashta settlements were on the banks of streams running through the steppe of the southern Urals. They were fortified with wood and stone constructions on earth banks by ditches, with houses often sharing walls. These were not city states; but they were social hierarchies, with social status indicated by the presence of horse and chariot in burials: chariot driving was for the few.[70]

The cemeteries of Kamennyi Ambar-5 and other Sintashta sites indicate that spoked-wheel chariots were first present in this region by around 2000 BC.[71] The Sintashta chariots had wheels typically about 120 cm apart, not very different from those of a cart or those which would develop later in the ancient civilisations of Egypt and China, although one example was 200 cm (80 in).[72] The single axle of Sintashta chariots had wheels with eight, ten or twelve spokes. The diameter of the wheels was very varied, from 74 to 121 cm (29 to 48 in). They were pulled by a pair of horses, whose harness equipment included what could now be confidently identified as cheek pieces.[73]

A dramatic example of a chariot was found in Grave 30 at Sintashta, where a man is buried between two horses. The wheels of the chariot they hauled, though decayed, are indicated by the depressions in the ground. Horse cheek pieces, a spear and arrowheads accompanied him.

At Krivoe Ozero in the south Urals, a chariot buried with humans and two horses was dated to approximately 2000 BC, so contemporary with the Sintashta chariots.[74] In the area between the Volga and the Don, horse bones were found in both burial and domestic contexts, and distinct carved antlers considered to be cheek pieces seem to confirm this.[75]

From 1950 BC onwards, samples of horse bones from archaeological sites in Bronze Age Europe indicate a significantly higher proportion of males than females, unlike the earlier periods of horse hunting.[76] This suggests that mares were maintained for their breeding value while male horses were slaughtered for meat – perhaps when too young or too old or too many to be needed for haulage or riding. During the second millennium BC, the horse-drawn chariot extended into the Hungarian grasslands and further into Bronze Age Europe.[77]

The major impact of the chariot was its adoption south of its area of origin by the urban civilisations of the Middle East, where its impact was more than just in technology and the status of owners; it would form part of military might. Actual horse remains are relatively rare in the archaeological excavations of the Middle East, where excavated examples of chariots are even rarer.

Horses had reached the region a little earlier than the chariot. There appear to have been wild horses killed for meat across the Caucasus region (between the Caspian and Black Seas) into Anatolia and further in the Levant.[78] The domesticated horse is mentioned in texts of the Sumerian Third Dynasty of Ur (ca. 2100–2000 BC) in southern Mesopotamia, apparently an exotic rarity kept in small numbers and in one case fed to lions as a public spectacle.[79] Both horses and the less powerful onagers were ridden.[80] A text from the period refers to a horse 'with flowing tail', distinguishing it from the onager or ass (donkey). An 18th-century BC text from Tell al-Rimah in northern Iraq implies a basic pack system in the phrase 'let the horses speedily bring the case of silver cups'. A burial with a four-wheeled wagon showed that donkeys were still a source of haulage. Images of horse and rider demonstrate horses were also present in Syria at about the same time.

A spoked wheel on a four-wheeled wagon rather than a chariot is shown on seal designs at Karum Kanesh, an Assyrian colony in Anatolia, dated around 1950 BC. This suggests the idea of lighter wheels spread a little independently of the association with the chariot.[81]

The chariot with its horses gave a new military strength from the 17th century BC to people who would challenge and even invade the lands ruled by the Mesopotamian powers or Egypt – Kassites, Hurrians, Hittites, and the enigmatic Hyksos – until the chariot became part of the equipment of these major powers themselves. The chariot itself saw refinements; Middle Eastern chariots typically had fewer spokes than in the north – four or six.[82]

South of the Caucasus in central Anatolia, a chariot drawn by two horses was depicted on seal impressions at Kültepe dating from about 1900 BC. The Hittite Empire which emerged from eastern Anatolia used horse and chariot by the 17th century BC. This augmented their military strength as they extended their power further west in Anatolia and into the northern Levant. They would clash with Egypt at the Battle of Qadesh in 1274 – by which time Egypt had also acquired chariots.

The Mitanni kingdom which controlled northern Mesopotamia and the northern Levant from ca. 1600 to 1200 BC used battle chariots. We have the text of a manual from a Mitanni horse trainer called Kikkuli in the service of the Hittites in the 15th century BC, giving detailed guidelines for the care, exercise and feeding of the horses required for the chariots.

The Kassites who controlled Babylonia (southern Mesopotamia) from about 1600 to 1150 BC also used the power of the chariot. The view that the best horses were now in Mesopotamia was reflected in texts from elsewhere. Correspondence on cuneiform tablets from Anatolia, north Syria and Mesopotamia emphasises the importance of the care and breeding of horses – with details of age, colour and sex.[83]

Rock art in the Sahara suggests that horses and chariots could possibly have crossed through desert regions in the mid 2nd millennium BC.[84] But Egypt was relatively late in the Middle East in using the horse-drawn chariot. The horse is generally thought to be an introduction into Egypt only in the Second Intermediate Period (1700–1550 BC), when the Hyksos, who used horses, established power in the land.

The remains of a horse at Buhen in Nubia, generally thought to be dated to the 17th century BC, provide an early example of use in ancient Egypt. A mummified horse in a coffin was found in the 15th-century tomb of Senmut at Thebes. Egyptian rulers would subsequently adopt the chariot with enthusiasm. Ten chariots from Egypt's New Kingdom date between 1550 and 1050 BC; by the reign of Ramesses IV (1155–1149 BC), his army had 50 chariots for every 5,000 infantry.[85]

Chariots and wheels from Armenia date to 1400–1300 BC; chariots would only reach China from 1200 BC.[86] Our information largely comes from images: in some such artwork, it is identifiably stallions which pull the chariots.[87]

There can be no doubt that the elites of ancient Middle Eastern states used the driving of chariots to emphasise their status in the streets of their town and travels through their domain. But the use of the chariot in warfare had changed the balance of power within and between Anatolia, Mesopotamia, parts of the Levant and then Egypt. It was a feature of successful armies until the Assyrians after the 9th century moved their military emphasis from the chariot to the mounted horseman and the flexibility and power of cavalry.

Chariots were a possession for display and status in Mycenaean Greece and in the Classical world, where some images show three horses harnessed to haul just a single man on his chariot. The Italian chariot was too narrow for an archer and driver, carrying just one individual. In Italy, their use is indicated by miniature bronze sculptures from the 11th to 10th century BC, and remains of chariots have been found from the 8th century BC.[88]

The Spread of Riding

When we think of the horse in human history, especially in the centuries immediately before mechanical transport, we often focus on the individual rider mounted on the saddle of their horse, ably guiding its speed and direction with reins. We may have a prejudice derived from our images of European or American history that the lone rider on their horse has a status above that of the mere passenger in a wheeled vehicle. That seems not to be the pattern of the ancient societies of the Old World, before the armed horseman came to dominate patterns of conquest and acquired a symbolism of secular power.

We noted previously that the horses ridden in hunting in the Central Asian Copper Age of Botai were not ancestral to the modern horse. We see horses tamed in the western steppe for herding and hunting, while oxen and donkeys were used to pull wheeled wagons. The change that probably began in the steppe and spread to the Middle Eastern civilisations was the use of horses not primarily to be ridden but to haul the innovation of the spoked-wheel chariot, where the horse could provide speed without needing the strength of the slow ox or the dependable donkey. So in the advanced civilisations outside the steppe, the high status of an individual owner and rider of a horse emerged *later* than the chariot. The elites of the advanced world initially wanted horses to pull chariots and for no other purpose. Studies of horse bones from Chinese chariot burials have shown the possibility of distinguishing in the archaeological record between horses for riding and horses for hauling chariots.[89]

Owners of chariots needed a groom to care for their horses and a knowledgeable person to train the horses to respond to commands. In Egypt, illustrations focus on the chariot; where an individual is riding on a horse, it is not a pharaoh or official but a servant or groom. In the two-man battle chariot, the warrior needed a driver he could trust. Being the rider of a horse was therefore not a high-status role.

The low prestige of riding is shown in an official missive to Zimri-Lim, king of the upper Mesopotamian state of Mari from ca. 1775 to 1761 BC: it advised that he should stop riding a horse if he wanted to impress subjects, as it would give a poor impression:

My lord should preserve his dignity ... Thus my lord should not ride horses, but a chariot with mules and maintain the prestige of his sovereignty.[90]

A tablet from the Sumerian Third Dynasty of Ur in the 2030s BC has a seal design which shows a rider on what appears to be a horse, yet there is no indication of harness; nor does the rider wear anything that might signify him as a person of status.[91] Baked clay plaques from Iraq dated to the earlier 2nd millennium BC show horses with riders, although similar scenes involve an onager instead. There is nothing to suggest that this was a privileged experience or associated with people of high rank.[92]

The gradual change in the status associated with riding the horse is seen in its use (or representation of its use) for hunting by the earlier 1st millennium BC. But the major change came in the Middle East with the Assyrian innovation of cavalry from the 9th century BC – the armed soldier on his own horse. The prestige associated with the horse as a symbol of military strength was largely what led to its identity as a useful and acceptable part of civilian life too. The owner of a horse was a man who could fight when his loyalties and affiliations required it. His use of his horse to travel in peace served as a reminder of his valour and his status. Ability to ride a horse, as indicated by Xenophon's *Peri hippikes* of the 350s BC, was a well-acknowledged skill in ancient Greece.

In Imperial Rome, it was a sign of status for a man to ride a horse in public.[93] The horseman of the later Classical world, the armed warrior of Islamic empires, the knight or landowner of medieval Europe, all linked civil and military roles and images. And the 'hordes' of armed warriors from the steppe who raided and even conquered settled civilisations represented something different still.

And to the East?

The main narrative of horse domestication comes from taming in the steppe societies and adaptation of the wheeled vehicle as the light two-wheeled horse-drawn chariot, enthusiastically taken up by the emerging and established states of the Middle East and then the Mediterranean. But what of the other end of the steppe, east to Mongolia (where the horse became central to the traditional culture of modern times), and elsewhere in Asia?

Domesticated plants and sheep, cattle and goats (but not horses) had been acquired by Bronze Age people in the mountain regions of Central

Asia by the middle 3rd millennium BC.[94] Studies of teeth from the Early Bronze Age indicate consumption of milk from cattle, sheep and goats, but not from horses.[95] The prosperous Bronze Age Oxus civilisation (the 'Bactria–Margiana Archaeological Complex') of ca. 2400–1900 BC lacked horses.

The horse seems to have come to the eastern parts of Central Asia together with the idea of the two-wheeled chariot and only then developed other uses. Evidence for consumption of mare's milk seems to arrive at the same time as evidence for riding – from 1200 BC. The adoption of the horse by human groups was marked by mobility, horse-meat consumption and use of horse bones in ritual. The chariot features in the early Vedic texts of India from ca. 1200 to 1000 BC, which in turn reflect a Central Asian (and Indo-European) background.

Research in Mongolia identified domestic horses in the Bronze Age Deer Stone-Khirigsuur (DSK) culture of the late 2nd millennium BC, where burial mounds were often surrounded by horse burials. Art images show chariots with a central yoke and two (or even four) horses. The horse skeletons indicate the use of bridles and veterinary care.[96] The dental wear raised the question of whether these horses were ridden or used only for pulling chariots: the position of the bit would have provided limited control for a rider, and chariot use seems the primary if not sole use of these domestic horses.[97]

On the steppe in the High Altai of the Siberian border with Mongolia, the Pazyryk Iron Age of the mid 1st millennium BC had saddles for riding horses and developed carriages with four spoken wheels and a plank roof held up on rod uprights.[98] However, to the south in the Tarim Basin (the Uighur region of China's Xinjiang), a human burial with a leather saddle, lead and horse skull dates from ca. 1000 BC.[99]

The horse and chariot enter dramatically into Chinese civilisation at a similar time as they are known from Mongolia, at about 1200 BC (although there have been some suggestions wild horses had reached there and were hunted earlier). The Shang dynasty which ruled from Anyang (Yinxu) is marked by the burial of both horses and chariots in pits. The chariot appears to have been used for ritual parade, for hunting and for warfare. By the Early Zhou period of 11th–10th century BC, a chariot drawn by four horses was in use, and the Zhou conquest of Shang China reflected both the quality of their horses and the design of their bridle bits to control them.[100]

Confronting the horse ridden for military use by China's neighbours on the steppe, China itself adopted the practice from the 4th century BC onwards.[101] Horse skeletons from Xinjiang dated about 350 BC show evidence of the impact of riding and of bits and bridles.[102]

The innovation of the stirrup was known by the 4th century AD in China and in Japan.[103] By contrast, in the 9th century AD elite Chinese court ladies were known to have ridden donkeys.[104] As China battled with the nomadic horsemen beyond its borders, such as the Xiongnu, these began to travel and raid west, in some interpretations being ancestral to the Huns who impacted the history of Europe in the 4th century AD.[105]

The innovations of horse taming and wheeled vehicles were revolutionary contributions to human societies wherever they reached. The post-Columbian transformation of the Americas, north and south – which lies outside the framework of this history – is a major demonstration of the difference which riders on horseback made to worlds where fast ridden domestic animals were previously unknown.

The Multiple Roles of Horses

The horse has played multiple roles in human history, affecting many different aspects of family, social, economic and political lives. The invention of the wheeled vehicle transformed the nature and speed of human movement. When the horse and the wheeled vehicle were brought together, life and work (and war) would be quite different. A full history of the horse in human society would fill a very long volume, but the early impacts of different stages of horse and wheel were dramatic and varied.

The taming of horses was different from the taming of fire: once fire was tamed, all its applications were available together. As the early history of the horse shows, in the complex (and vigorously debated) stages of its domestication, the horse played different roles in different contexts.

Our own popular sense of the horse in the human past is influenced and perhaps dominated by some stereotypical images. One is from North America: the cowboy in the open spaces and the Native American hunting bison on the plains and fighting from horseback. A second impression is from literature and the relatively modern history of the Western world. This presents every man of authority riding his own horse in the countryside and from town to town, while carriages conveyed the ladies and their children, stagecoaches made the long-distance journeys, and horse-drawn hansom cabs moved people around their city appointments. A third image brings us the steppes of Asia and the world of traditional pastoralists, the focus of whose lives and mobility relied on their horses; and a world whose armies of mounted warriors influenced history east and west.

The Horse for Meat

Meat from horses and their close relatives has been important in a large range of societies, from Ice Age Europe to modern France, from the cold winters of the Eurasian steppe to those hunting wild asses, onagers or zebras elsewhere.[106] The paintings of horses in Upper Palaeolithic Europe reflect this significance.

As the vegetation of Europe changed at the end of the Ice Age, horses became a smaller part of the wild food resources available to forager groups. The limited variety of game animals in grassland areas such as the steppe made human reliance on horse meat a necessity there. The hunting of wild horses for their meat, a major part of the economy of Copper Age people of areas such as Botai in Kazakhstan, continued to supplement the diet of settled farmers in the western steppe.

There in the steppes, the ability of the horse to forage in the frozen winter gave it a wide distribution. This continued as the horse was domesticated and became a core possession of nomadic pastoralists up to modern times, in and beyond Mongolia and northern Kazakhstan. The high dietary value and palatability of horse fat is especially appreciated in these areas, and the contribution of protein and fat from horses can compensate for the lack of other regular meat sources.

In the traditions of Kazakhstan and Mongolia, meat is smoked to be consumed safely later, and meat from horses slaughtered in November can be kept frozen over the winter. Kazakhs traditionally slaughter surplus young animals between one and four years and older animals who may be aged from fifteen up to thirty, leaving those stallions fit for riding and females to breed and provide milk.[107]

Cultural factors have affected the consumption of horse meat in other areas where horses are kept for riding and haulage. Traditions or formal prohibitions against the consumption of horse meat occurred in contexts where there were many alternative options. In times of war or shortage, such prohibitions might be expected to change, as in Second World War Britain.

Dietary laws in the Jewish community have been interpreted as prohibiting the eating of horse meat. The suitability of horses for human consumption was debated by medieval Muslim scholars, and while it has not been classified as forbidden, horse eating is rare in most Islamic societies.

The early German pagan habit of consuming horse meat raised issues for Catholicism seeking to extend its authority in the 8th century; Pope Gregory III described it as 'a filthy and abominable custom' in 732,

The Multiple Roles of Horses

effectively declaring a position on the consumption of horse meat held by Catholics until the 18th century.[108] France is known for its horse-meat butchers' shops, and today, outside of the lands in and bordering the steppe, horse-meat production is significant in Mexico.[109] In countries where the market for horse meat is small, it is exported or made into pet food.

The Horse for Milk

Within communities of the modern steppe, groups who consume horse meat from their domesticated stock also typically acquire mare's milk. In archaeological studies of prehistoric steppe peoples, the search for fats from mare's milk on the surface of pottery can help clarify the sequence of domestication, since while meat can come from wild or tamed animals, milk is only available from a mare retained in captivity and with young.[110]

In traditional Kazakh societies, horse milk was primarily obtained in summer from a mare with an unweaned foal, and in its raw form it is fed only to human babies, for whom it serves as a weaning food.[111] Raw milk has a laxative effect on adults.

Adults only consume horse milk in fermented form (*kumys*/*koumiss*, also known in Mongolia as *airag*), which has been a staple food across the steppes of Central Asia and beyond. Traditionally, it was used to give strength, being taken before a long journey or carried for consumption during it. The process of fermenting can be rapid – churned evening and morning, it is then ready to drink. Like horse meat, it is high in polyunsaturated fats, and a good dietary alternative to cow's milk, with many vitamins; the alcohol content from the fermentation is low. Today, *kumis* is available bottled commercially.

Secondary Products

Discussions of secondary products obtained from early domestication often focus on wool from sheep or dairy products from cattle or goats. But from the horse comes another important secondary product: hides. The horse hide can be used for leather, today especially in shoes. When Stone Age hunter-gatherers were hunting horses in Europe, Asia and North America, we can assume that the hides of horses were valued after the meat was consumed. Among the steppe horse hunters and herders, the straps and equipment for controlling a horse were made from horse leather.[112] Because of the decay of organic materials, its presence or absence in prehistoric communities is hard to test, but it is unrealistic to think that the hide of a slaughtered horse was not available for multiple practical uses. The bones of horses, like those of other animals, could be used as a raw material to carve.

Riding to Hunt

The evidence from Botai in Kazakhstan suggests that the first domesticated horses were tamed by a community dependent on hunting wild horses for their meat. Hunting horses is no easy task. If they can be driven into an enclosed space by a group of hunters – a valley end or corral – they can be speared, but a herd of wild horses, whether bachelor males or a group of mares led by a stallion, could easily outrun the hunters and keep up their speed far longer. If healthy younger men can run for a period at something like 10 km or 6 miles per hour, a wild horse can canter at double that or gallop at three or four times the human pace – and maintain that speed. The horse was not the easiest prey of those represented on European Palaeolithic rock paintings. But once the horse hunters of the steppe who relied on horse meat had raised horses in captivity, trained them to carry a man and harnessed them to keep the rider secure and the horse subject to guidance, then the hunter and the prey had much more speed and endurance.

The mobility and speed of humans on horses expanded the role of the horse in early civilisations. When the animal's role was to haul the two-wheeled chariot of a member of the civilian or military elite, only junior staff would need to ride it for training or stabling. But when those same members of the social elite went to hunt wild animals – not for food but for sport and for status – they would become the riders themselves, though the practice was not seen before the 1st millennium BC.

This new role is visible on the relief of the lion hunt of Ashurbanipal, king of Assyria from 668 to 631 BC (Figure 10). This was erected on the wall of his palace at Nineveh in northern Iraq and is now in the British Museum. His expedition includes a chariot, but the attack on the lion shows the king sitting on the cloth saddle of a harnessed horse. By this time, the Assyrians had moved from chariots in war to armed cavalry. So there was a long gap between riding by Botai meat eaters and riding by Middle Eastern elites.

Perhaps the most dramatic impact of horses used for hunting was when the horse was introduced to the Americas, and particularly when horses spread north, initially by trade from Spanish settlements to the Native Americans of the Great Plains and the northern Rockies, where horses preceded human colonisation, then by local

The Multiple Roles of Horses

FIGURE 10 Assyrian king Ashurbanipal hunting on horseback (Carole Raddato, CC BY-SA 2.0, via Wikimedia Commons).

breeding. While historical data had suggested this was a feature of the century from 1680, genetic studies have shown the earlier spread into the Great Plains and Rockies in the first half of the 17th century.[113]

'For the first time in the region's long history men and women were not limited by their own speed and endurance ... the horse rearranged the basic alignment among people and the possibilities that flowed through the plains.'[114] Possession of the horse transformed inter-group warfare, but it transformed the economy more as the importance of farming gave way to the potential which horses provided for hunting game across the plains – primarily bison – and especially when the rifle was also traded. Horses did not compete with their owners for food; the grasses of the high plains provided year-round sustenance.

Riding to Herd

The mixed agriculturalists of much of prehistoric Europe and much of prehistoric Africa typically grew crops in their fields, had a small number of domestic cattle, sheep, goats, pigs and fowl and lived in farmsteads of small villages accessible from their lands. Communities in the open grasslands of the steppe with large animal herds needed means to manage and control these. Over the millennia of pastoralist occupation, they might need to keep their animal herds grazing together, corral them for safety and move them to upland pasture in summer and back for winter.

A shepherd on foot may be able to control as many as 150–200 sheep. If he has a horse, he can control as many as 500.[115] The horse-riding shepherd or cattle herder can bring in an errant individual, check out grazing land ahead and move quickly round their flock. The horses whose bones we find in settlements of earlier mixed agriculturalists of the western steppe could have served a herder in such a role.

From about the beginning of the 1st millennium BC, communities that had no crops could survive from pastoralism alone, trading with settled farmers and moving larger herds around the grassland with the aid of horses. The Mongolian horse riders of modern times were primarily riding geldings; traditional Mongolian horses were broken at two to three years, then ridden from three to four years.[116]

The cowboys on great cattle ranches of the Americas, north and south, and later the stockmen on the vast sheep and cattle stations of Australia, depended on the mobility provided by the horse. A man might determine when and where to move stock, a dog might assist in rounding them up, but the horse gave a practical assistance that made the stock-based economy of large landholdings work.

Riding to Communicate

Traders dealing in goods through prehistoric agriculturalist Europe or the western steppe not only spread the exchange of raw materials and products but also served in the spread of ideas. It was not large wagons that were traded but the idea of constructing them; the spoked wheel was an idea to copy not an object to buy. Gradual communication inspired social and technological change.

But in the worlds of competing empires to the south, as to the east, a different kind of communication was of vital importance. The administration of the state relied upon information and instructions travelling between the central ruler and outlying districts. The pharaonic Egyptian state had the domestic communication advantage of the Nile along which boats could convey messages and messengers. Mesopotamian states could also make use of their rivers.

The 'Royal Road' under the Persian Achaemenid rulers of the 6th–4th centuries BC allowed fast communication across much of a vast empire. It was said that the 2,700 km (1,700 miles) were just nine days'

The Multiple Roles of Horses

journey for a sequence of riders, an achievement that left Greek historian Herodotus in awe.

Military strength required commanders to maintain contact with forces some distance away. Intelligence was needed to report military developments and threats which might emerge from a rival empire. Winning a war required speed, although signalling by fire beacons could help. A runner carrying a message might serve the purpose, but a courier by horseback was significantly faster. This could also provide more security: an armed man on a horse might be less open to attack and more able to defend himself than a runner on foot.

And to keep the peace, diplomatic correspondence was needed between rival states. Diplomatic exchanges between rulers might mean a formal expedition on foot – including an escort from the recipient court. Messengers travelling by horse significantly cut the time to communicate a message to a distant ruler, although the delays in being detained at their court extended the overall time frame of an essential exchange.[117]

Gradually, couriers on horseback took on the role of diplomatic or administrative messengers with increasing efficiency from the 1st millennium BC. But not always: the Olympic marathon distance of 42 km (26 miles) is based on the foot journey allegedly run by Pheidippides to report to Athens a military victory in 490 BC. A major visible impact of the Roman Empire was, of course, the networks of constructed roads. These not only facilitated trade, administration and military movements but also provided a basis for rapid transmission of official messages at the speed of horses.

The Mongols of the 13th and 14th centuries AD, with their origins on the steppe, were of course masters of the horse. In Genghis Khan's empire, a relay system used regular changes of horses which in open areas meant a message could travel 200 km (125 miles) a day – by some estimates even more. This system expanded further during the Mongols' rule of China.

State administrations as well as individual aristocrats in Europe, as in Africa and the Middle East, could send their own personal couriers on horseback to deliver a message. With the rise of a middle class in Europe and the complexities of business life, the horse-drawn mail coach developed to take people and the mail together along regular routes and from the 1780s operated between major cities in Britain. The short-lived Pony Express in the United States of 1860–1861 was badly timed, since the electrical telegraph was soon available for quick messages and the railway for other letters. In Europe, the railways superseded horses and horse-drawn coaches from the mid-19th century, bringing to a close three millennia of equid-led communication.

Riding to Fight

Before formal military use of the horse and rider as recorded in ancient Middle Eastern states, it is very possible that in the steppes the ability to ride a horse served communities used the ability to ride a horse when they were in conflict with their neighbours.[118] However, the advantages of speed and height provided to a warrior by his horse could be offset by the limits to flexibility when survival depended on the simultaneous control of both horse and weapons. The development of horse-riding military forces – cavalry – would appear to be a major driver in world history for the development of other military technologies and the social and political impacts these brought about.[119]

The core fighter in the civilisations of the ancient Middle East fought on foot, whether with sword or spear or as an archer. Whereas conflict in smaller polities of agricultural Europe, Asia or Africa is likely to have involved many of society's young men, it was professional armies that gave the rulers of Middle Eastern states their strength. The horse-drawn chariot began to change the balance of power, but that was the weapon of a minority.

The idea of the man on his own horse being armed and ready to fight may have developed on the steppe, but the change to the use of cavalry in the civilisations of the Middle East seems to have emerged only by the 9th century and in the Mesopotamian state of Assyria. Shalmaneser III, the Assyrian ruler from 859 to 824 BC, is shown riding on a horse himself. Before that time, the Assyrians were fighting with chariots; they then developed cavalry, and by the 7th century horse-riding archers had replaced the chariot.[120] The horse provided a high position which allowed an archer to shoot without needing to engage closely to an enemy – but unlike the two-man chariot with driver, the horse-riding archer had to control his own horse. In the initial stage of the mounted archer, as with the chariot, two horses rode together, the rider of one leading the reins of the other, whose rider was then free to fire his bow. Archery was also, of course invaluable in hunting; the frieze from Nineveh now held at the British Museum shows horse mounted bowmen hunting (Figure 11).

Militaristic nomadic pastoralism is exemplified by the Cimmerians from north of the Black Sea who raided south using their strength as horse riders, and the Scythians who were instrumental alongside the Medes from Iran in bringing about the end of the Assyrian Empire. Their facility with horses was significantly helped by the design they developed for bridle and reins, which aided difficult manoeuvres.[121]

The Multiple Roles of Horses 77

FIGURE 11 Assyrian archers (Osama Amin, CC BY-SA 4.0, via Wikimedia Commons).

The Medes could attribute much of their success to their horsemanship, as did their successors, the Achaemenids. These controlled a vast empire until in turn it was defeated in 334–331 BC by Alexander the Great, whose horse Bucephalus has become a figure in history by himself. The later Parthians from northern Iran used heavy armour to protect both horse and rider.

It may be that in western Europe the riding of horses only dates from the 7th century BC, driven by military needs.[122] Mounted horsemen featured in Italian imagery by the 6th century BC. Hallstatt Iron Age warriors rode horses while they fought with a slashing sword. Once an enemy had acquired this weapon, others would need to follow.

Earlier on in the steppes, the development of armed warriors in command of their own horses became a threat to imperial China. In China, the chariot was used in warfare before the appearance there of the cavalryman. Chinese imperial power was threatened periodically (and sometimes very effectively) by horse-riding warriors from the steppes to their west. China acquired horses for military and other uses, but the stronger, more powerful horses were bred by their enemies. In 104 and 102 BC, the Chinese sent expeditions west with a particular aim of seizing horses from the steppe.[123]

Armed steppe horsemen would threaten societies to the west and south too. The most successful of these, the Mongol Empire, would end the Abbasid rule of the Islamic world in 1258 and reach into eastern Europe

FIGURE 12 Mounted archer from the Mongol army (photo credit: CPA Media Pte Ltd/ Alamy Stock Photo).

(Figure 12). One great advantage of the army of horsemen lay in their speed, enabling them to take an enemy by surprise: raiding and attacking without warning or alerting an enemy's intelligence and if necessary moving away again just as quickly. In some campaigns, one warrior might take with him five horses to allow him a change or to replace a wounded horse. An estimate suggests that while an infantry force might move 25 km (15 miles) a day, and US cavalry on normal patrol with equipment and would travel double that, a force like the Mongols' could manage 110 km (70 miles).[124]

The stirrup allowed the most effective military use of a horse. A cavalryman held safely in position by a stirrup was less likely to be dismounted by a body blow (or be thrown by an alarmed or wounded horse). The stirrup assisted him to manoeuvre fully with sword, pike or lance without the danger that the force of his own blow would unseat him.[125] An archer who was an accomplished horsemen using the security of a stirrup would be able to fire arrows in all directions, even behind him when retreating, and this gave armies like the Mongols a great advantage. Stirrups to assist in mounting a horse may have been present earlier in

The Multiple Roles of Horses

India, but the pair of stirrups for riding was known in China by the 4th century AD. It may be from there that its use spread westwards into the steppe and also into Siberia. Introduced to Europe in the 7th century, by the 10th century the stirrup was established among European riders and essential to the medieval knight.[126]

Cavalry developed in importance throughout the Old World, and in the armies of both colonisers and Native Americans in the New World. While armies might still numerically be dominated by men on foot, those in charge of battle strategy would be mounted. 'My kingdom for a horse', cries Richard III in Shakespeare's play when he has lost his mount. The armed and armoured horseman remained a dominant force of European empires until the development of artillery in the 16th century.

The cavalry charge is a feature of our historical imagining (and its film versions). The cavalry remained active in the First World War: Australian horsemen took the last successful cavalry action at Beersheba in Palestine in October 1917; and cavalry took part in operations in the Second World War in Russia and Poland – 2,800 years of horsemen in battle. And of course the ceremonial symbolic role of a military leader on horseback remains.

Horse and Chariot

The emergence of the vehicle with two spoked wheels, hauled by two horses and carrying one man or two, allowed for speed impossible with a heavy wheel or a vehicle pulled by oxen or donkey. In peace, such a chariot would serve as a symbol of authority and prestige. In war, with a driver and a military figure such as an archer, it could play a powerful role in battle.

Nevertheless, the chariot had its limitations in warfare. An infantry soldier had greater control of his movements and the use of his weapon than did a charioteer. The chariot could mark status, and its speed allowed an officer to move rapidly around his troops. A charging chariot might break up a group of infantry, and a row of chariots driven towards enemy foot soldiers could cause fear, but once the chariots had reached and passed through enemy lines, they had limited manoeuvrability.

In the two-man chariot, one rider could control the horse while a second man could wield a slashing sword.[127] With his companion directing the horse, an archer could fire arrows from added height and distance and move to do so from different positions. Chariots would also be used in sieges to patrol and to surround a besieged location.

FIGURE 13 Pharaoh Ramesses II and his chariot (photo credit: colaimages/Alamy Stock Photo).

In pharaonic Egypt, in addition to images on temple walls of the pharaoh in a chariot (Figure 13), tombs of elite officials included visual presentation in highly stylised form of the owner on a chariot.[128] More realistic images of horses are found in other contemporary Egyptian contexts.

A further weakness of the chariot as a fighting machine was that a wound to the horse would disorient it or bring the chariot down. Even after the use of scale armour on a horse, which had already been present in the Late Bronze Age, there was more flesh for a foot soldier to strike on a horse than on its better-protected rider.[129]

The *kurgan* graves of the Bronze Age of Eurasia beyond the territories of the powerful states contained not just horses and chariots but the weapons used by chariot riders: spears and arrows. The ability of a man to fight or hunt while using a chariot requires much practice; a single rider might wrap reins around himself while shooting missiles but not be able simultaneously to guide his horse's direction. A chariot has to give advantages beyond those of riding the horse directly; the individual mounted warrior unencumbered by wheels made the armed peoples of the steppe a dramatic force.

The Multiple Roles of Horses

Racing the Horse and Chariot

The use of chariots in warfare inspired its appearances in sport. Horse racing would follow, a feature of the Classical world. Both horse racing and chariot racing were part of the Olympic Games by 648 BC.[130] In ancient Roman displays, *hippika gymnasia* allowed skilled horsemen to show off their skills in a competitive environment. The horse races in Rome's Circus Maximus were on a course 650 m long.

Versions of polo are found in many parts of the Old World, with its early development in the Persia of the Parthians (247 BC–224 AD) and Sasanians (224–651), and it remained a feature of Persian society under Islamic rule. It spread eastwards to India and China and westwards to Constantinople, where it was known in the 4th century.[131] The game called *buzkashi* from Central Asia is even more dramatic, involving horsemen and a goat carcass, not a ball, and the game with a ball played by the ancient Persian elite may have been derived from a more common version among the nomadic peoples of their Central Asian neighbours.

There was a military angle to competitive horsemanship in the courts of medieval Europe, famously involving also royal competitors in jousting and the like. This took on a more civilian air by the 18th century, and the range of sporting competitions involving the horse expanded to include a large number of competitors and the larger number of those gambling on the results. Steeplechases, show jumping, flat racing and trotting with the manned two-wheel cart have all moved the horse in sport from military to secular to commercial contexts. Today's Olympic Games include dressage and jumping competitions, and eventing to combine dressage and show jumping with cross-country. The Spanish Riding School in Austria presents dressage not as competition but as display.

The Pack Horse

A pack animal can serve to deliver goods where wheeled vehicles cannot be used – rough or mountainous or waterlogged routes. It can also carry produce for those who do not have the means to possess a cart or wagon. On the steppe, horses may have been the beast of burden themselves before they were adapted to pull carts and wagons.

Horses as beasts of burden lack the efficiency of other domesticated animals but nevertheless developed this role in appropriate locations over the millennia. A pack horse required panniers strapped across its back, and harnessing for it to be led either by a person on foot or by a horse-riding companion.

One incentive would be to carry supplies required by horse riders on their travels, since a pack horse will travel at the same speed as its companion rider on their own horse. A common image in modern history is of the explorer or trader riding through rough and perhaps unfamiliar territory accompanied by one or more pack horses. The pack horse was also a military strength, accompanying cavalry with supplies.

The donkey long retained its role as a pack animal in much of the Mediterranean region and expanded its use into Europe, Asia and Africa. A favoured alternative was a mule (the cross between male donkey and female horse). Mules and donkeys are considered more sure-footed than horses on difficult ground, with a temperament more suited to the role, and cheaper to feed than horses.

By the early Middle Ages in many areas of Europe, the pack horse had gained precedence over the donkey for carrying loads. The pack horse remained common in England into the 17th century.[132] Pack trains of mules or horses were a feature of European expansion westwards in North America, and donkeys found favour in Central and South America.

Transport by pack horse or donkey, cart or wagon, gave way to the dromedary camel in much of the Middle East and North Africa between the 3rd and 6th centuries AD, with the Bactrian camel used in Central Asia. The ability of the camel to carry loads over long distances in arid conditions gave it major advantages. Although first domesticated by the 2nd millennium BC, its importance in the Middle East (including Egypt and Assyria) grew from the mid 1st millennium BC, probably led initially by traders from outside those major civilisations.[133]

Meanwhile in South America, where transport did not involve wheeled vehicles, the domesticated camelids served as pack animals, with the Incas' llamas hauling loads both locally and on the complex of long-distance roads.

The Horse for Haulage of Goods

The history presented earlier in this chapter emphasised the separate development of the wheeled wagon and of horse domestication. In its origins, the wagon for hauling goods was most effectively pulled by a pair of oxen, and in the Middle East by donkeys. Oxen offer in strength what they lack in speed. It was the chariot, not the cart, which used the horse and wheel together to achieve faster pace. An ox hauling a cart or wagon might travel at little more than 3 km an hour; a horse with a chariot could reach speeds almost ten times that.[134]

The Multiple Roles of Horses

When horses and confidence in their handling became commoner, the horse gradually replaced oxen in many places. The choice of animals to pull carts or wagons was influenced by cultural context, available stock and individuals' economic position.

Even in the world of Classical civilisations, the horse was not the norm for haulage. In Italy, wheels are attested from the 2nd millennium BC, and burials of carts and wagons have been recovered from the 8th century BC onwards. Mules were favoured for haulage from the 6th century BC; indeed, in Classical Greece and Italy oxen were less in favour than mules or donkeys.[135] The two-wheeled cart was more common than the four-wheeled wagon in representations until the period of the Roman Empire. Horses came to be used as an alternative form of haulage in the Roman world, but their use declined thereafter. Horses became more popular again for haulage in Europe much later, in England only beginning to replace oxen by the 12th and 13th century, with an estimate that around the year 1200 only 5 per cent of draught animals in England were horses. The 16th-century adoption of four-wheeled wagons with several horses hauling a heavy load put bridges at sufficient risk that the authorities sought to outlaw them in favour of two-wheeled carts.[136]

Horses pulling barges on waterways could haul much larger weights than on earth tracks. This was a feature of ancient river use but of course became of crucial importance with the development of the modern canal networks of Europe in the 18th century. By then, horses were also replacing oxen to pull an agricultural plough.

Yet today, the term 'horsepower' is still the measure of strength, established by James Watt to define the output of a steam engine. It was applied to other forms of power, though still using the formula devised by Watt and based on the assumption that a horse could turn a 12-foot (3.7 m) radius mill wheel 144 times an hour.

Turning a mill wheel (for which examples date as early as Greece in 300 BC) was yet another practical use of the horse in human history and an alternative to using the power of water or wind. But it was an alternative that requires human control to harness, lead, feed and stable the horse, unlike the natural power of water and wind. And horses, of course, were used in underground mining from the 18th century.

Transport around cities remained dominated by horse-drawn carts and wagons well into the 20th century, in roles as diverse as garbage and night-soil (sanitation) removal, taking domestic supplies to and from shops, and cartage from docks to warehouses and railyards. The 300,000 horses of London in 1893 was perhaps the high point in their use and represents a remarkable figure and a remarkable diversity of applications.[137]

The Horse for Haulage of People

The use of a horse to pull humans in a two-wheeled chariot was a specialist role, applied primarily in a context of demonstrating status, for action in war or in competitive sport. Haulage of goods in a wagon or cart would follow when preferred to oxen or donkey or mule. But the use of horses to transport people in a vehicle was later still: the carriage developed in different contexts.

Early examples were simply a covered framework on a wagon to support a cloth that protected the travellers from wind or from view. Copper models from the Indus Valley civilisation sites of Harappa and Chanhu-daro show covered carts dating before 2000 BC.[138] In China, a two-wheeled chariot with a parasol became a passenger vehicle from the 1st century BC.[139]

In Europe, the carriage developed from a cloth stretched across a curved frame towards the solid wood construction of a carriage with sides. The covered carriage, the *harmamaxa*, appears in Greece by the 8th century BC, used especially for women's transport. People carried in open carts were featured in artistic representations on Greek pottery, including what appears to be a wedding party.[140] While rare, a covered carriage was used in Classical Roman contexts. Medieval European conveyances followed the model of a wagon with a curved frame, with a minority for the elite constructed with solid sides (Figure 14).

The century from 1550 to 1650 saw valuable developments of the carriage in Europe, with improved suspension increasing its suitability and use for long-distance transport. It would establish itself by the 19th century as suitable for carrying the old or sick, adult females or children, or the higher clergy, while men with the means rode on horseback. The horse-drawn omnibus served shorter-distance urban journeys with multiple passengers in Paris and London from the 1820s. In 1875, some 1,000 horses were owned by the London Omnibus Company; twenty years later, this had grown to 32,000, outnumbering the city's horses for cabs.[141] It was the railway that marked the beginning of the end of the horse-drawn carriage, but with only gradual effect.

Riding to Travel

Perhaps surprisingly, then, in much of the world the use of a horse to enable its owner to travel from A to B whenever required came much later than other uses. The horse had been used to haul a chariot; and then, instead of an ox or donkey, applied to pull a wagon or cart. Its status rose with horses

FIGURE 14 Covered carriage with women passengers, 1411, from the Toggenburg Bible (photo credit: akg-images).

trained to carry warriors into battle, and the use of a horse for individual transport developed.

Ownership of a horse lay with the wealthier members of society in the ancient world of the Middle East, the Classical civilisations and the Islamic empires. Those who could afford to own a horse were defined as the second most wealthy class in Solon's Athenian constitution of the 6th century BC. A statue of a Roman emperor on a horse could emphasise his power, authority and military leadership. In medieval Europe, where the aristocracy were expected to fight when required, the horse was a symbol of military prowess and status much as the chariot had once been.

With the transition to the early modern world, the wealthier sectors of society had an increasing incentive to travel without goods from place to place. Wealthy landowners needed to travel between and around their estates, administrators were required to move around as part of their work and for commerce people needed to travel from home to places of work. The court of Henry VIII had 199 horses, cared for by 98 men.[142]

The privately owned horse available for work or pleasure is only one of the services provided by its taming, and one that applied to a minority of times, places and people in the whole history of human relations with horses. Nomads of the steppes might rely on their horses for their economic lives; military status could be exhibited by officers on horseback or a body of cavalry. But civilian possession and use of a horse to ride were elsewhere available to only select classes through into modern times.

Until the 19th century, the horse and its rider were the means by which the upper and middle classes went about their business in Europe. Those who occupied large areas in the Americas to raise cattle or sheep relied on horses to manage their stock. Native Americans adopted the horse to hunt on the plains. In 1900, there were about 24 million horses in the United States.[143] All these could trace their tradition back to the steppes of the mid 4th millennium BC.

Horses in Ritual and Symbolism

The horse existed in people's minds as well as people's lives. Its representation in European Palaeolithic rock art went well beyond its economic contribution. The *kurgans* of the Eurasian steppe held horse bones buried with the dead, and horse sacrifices appear in other prehistoric contexts. After about 1200 BC, horses (commonly showing evidence that they had been bridled and ridden) were sacrificed and buried in substantial numbers right across the steppe of the Late Bronze Age, notably in Mongolia but spreading from southern Russia to north-western China.[144]

The importance of the horse in the European imagination is indicated dramatically by the prehistoric horse outline cut in the chalk of Uffington in England, 110 m (360 ft) across and dated to the late 2nd or early 1st millennium BC. Horse burials are found throughout Europe, and especially north-western Europe, as late as the 1st millennium AD.[145] While a few were linked to human burials, others are most likely associated with ritual feasts after which the bones of the consumed horse were ceremonially interred.

The importance of a horse gifted as a symbol of friendship or alliance can be traced back as early as the ancient Middle East, when the horse-riding Kassite rulers of Babylonia sent gifts of horse to the Egyptian pharaoh.

And a final note of how horses were more than just a practical tool. The archaeological finds of many model horses we can interpret as children's toys through the ages serve to remind us how important a symbol, and reality, is this possession of human societies.

Conclusion

Innovations in human society can begin for one purpose but extend to fit, or create, other functions. The ability to tame, control and breed the horse was such a development. The wheeled cart or wagon originated (probably inspired by use of sleds on rollers) to move goods, when oxen or donkeys could haul weights beyond the ability of their human owners. Once the horse and the wheel were brought together as the chariot, elites in peace and war had new tools. It would take some time before the solitary rider on his horse, or the horse-drawn carriage conveying a group of people, gradually spread through different social classes and different regions of the world.

It is today easy to forget how basic were the historical transformations brought about by horses to human movement and activity in the past, and how much we take for granted the speed and facility of the horse. But the horse is there when we read of a Roman senator moving between his estate and the city, or a general inspecting his troops; a medieval official collecting taxes, or an Asian or African ruler sending messages of peace or war to another state; or merchants travelling along lengthy Old World trade routes or European colonists arriving in the Americas.

The domestication of the horse contrasted with that of other domestic animals. For those societies and those individuals who controlled horses, their lives were no longer limited by human speed. For those who fastened haulage animals to wheeled vehicles, they were no longer limited by human strength. And the combination of horse and vehicle – whether chariot, wagon or carriage – would come to mark economic power for millennia throughout Old World societies.

If horses freed humankind from environmental limitations in these ways, other forms of liberation awaited. Horses might allow more rapid communication across space, especially valuable to emerging and expanding empires, but the invention of quite different forms of communication provided the elites of some such societies with a new tool – writing – and it would be innovations in communication which would underlie the construction of the modern world.

CHAPTER 4

Developing Writing

> Writing is of that efficacy as to make itself intelligible to all the powers and faculties of the soul, notwithstanding its being mute and void of motion. So charming a delicacy shines in a masterly curious hand, as creates in the judicious beholder, pleasure ineffable. Something there is that gives life and spirit to a letter, that makes strokes seem to move, and casts a kind of glory round 'em ... If the invention of writing be so wonderful, the use and ends of it so extensive, and the improvements of it so delicate; then, how much are we beholden to those gentlemen whose happy genius and publick spirit led them to scatter roses in our way, and to make the paths of science pleasant to us?
>
> Robert More, *On the First Invention of Writing: An essay*, 1716

The Origins of Writing

We often think of ourselves as in world dominated by the visual. Our smartphones capture, transmit and receive multiple photographs; our televisions and computers stream moving images into our homes; advertising posters display visual presentations of desirable products in our streets. But the written word remains powerful. On every short road journey, we confront words to give us directions, words to give us instructions, words to give us advice. Even the packet containing my morning breakfast cereal carries over 600 words.

This chapter describes the many and different contributions of the technology of writing to society, and the force and impacts achieved with the revolutionary innovation. Writing was a catalyst to changes in economic, political, social and religious relationships. The presence of a literate sector of society meant the impacts of writing were diverse, if ever developing and changing, and we are able to identify in this chapter some of these substantial transformative functions of writing in the societies which were early adopters.

Writing first appeared with the beginnings of urban civilisation and the emergence of the state in Mesopotamia (ancient Iraq) and Egypt's Nile Valley at the end of the 4th millennium BC. It would come to serve multiple roles with substantial and quite diverse impacts. The first societies we describe as civilisations, with new relationships of wealth and authority, developed symbols to represent sounds: a technique to convey a variety of messages and information in a form that could be widely understood across space and time. Increasingly complex economic relations required new means of recording numbers and measures and agreements on produce and land. Written signs spelled out names which identified ownership of goods and described activities of rulers, in Sumerian cuneiform script or Egyptian hieroglyphic and hieratic. Scribal schools practised the skill of writing by copying word lists and provided a stimulus for the creation of literary works in written form. Rulers used the written word as propaganda, proclaiming their achievements for their own citizens and their relationship with the gods. Spells and texts addressed to the gods could seek divine favours. Written language needed to be standardised to be understood across the administration of a state, and such a common language served to unify political entities. With writing, law could be codified, legal cases and agreements recorded. Kings could exchange letters with other rulers, while faraway military and civilian officials could communicate with home. Writing thus developed into a tool of power and the powerful in society.

The transformation of visual symbols from representations of *things* into representations of *sounds* is what we call the invention of writing. It has a different history from many other innovations, since the idea developed independently in a small number of places. This makes the innovation more comparable to pottery or plant domestication than to horse taming or wireless technology.

A symbol clearly representing a known subject (as in a cave painting or rock carving) may be able to reach audiences irrespective of their spoken language, though the meaning they read into that symbol may be quite varied. But writing can only be understood by those trained in its interpretation: that is, enabled to read the sounds indicated by the signs that have been written. The actual words indicated by the writing are specific to each language. A feature of much early writing is the absence of breaks between words, so the reader must interpose these for understanding; and another feature (as in Egyptian hieroglyphic and hieratic texts) may be the absence of vowels between the pronounced consonants, so that reading aloud required a further interpretation of the written form.

The emergence of writing forms a conventional break in the traditional division between 'prehistory' and 'history'. Its beginnings have long been seen as associated with the early stages of urban civilisation in the ancient Near East (within the Middle East of contemporary usage): Egypt (the lands of the Nile Valley and Nile Delta) and Sumer in southern Mesopotamia, spanning the Euphrates and Tigris river systems of Iraq.

The great and influential archaeologist Vere Gordon Childe popularised the phrase 'the Urban Revolution' in his writings from the 1930s, seeing writing as a natural component of the developments which marked the change from agricultural village settlement to town and state.[1] Childe's model saw human history marked by a Neolithic Revolution and an Urban Revolution (both initially in the Near East) comparable to the Industrial Revolution of the 19th century. The Neolithic Revolution was defined as the transition from hunting and gathering to domesticated plants and animals, a process we now see as more nuanced, regionally variable and complex than in Childe's model. The beginnings of urban civilisation, and the beginnings of writing, were marked by a shorter time frame.

The huge diversity of settlements we might describe as urban, towns or cities (and indeed the widely different use of the term 'civilisation') has shown limitations to Childe's overall model of historical development.[2] But it is still valid to write of a transition within areas of the Near East to state societies with concentrated populations, and the term Urban Revolution provides a useful shorthand for that specific context.

Correlation and association do not demonstrate causation, and debates continue on the nature and process of urbanisation and the origins of state societies. The relationship between the development of writing and the development of the state is a complex one.[3] It is unrealistic to think of writing as necessary in shaping the state – state societies in China, the Americas and elsewhere developed in complexity without the tool of writing – but writing once invented served to aid and emphasise the power and facility of state institutions.

Underlying the major shift in the ancient Near East was the development of agricultural surpluses. Production in excess of that required to feed a family could be traded for goods produced by those specialising in manufacturing and trading other materials, including tools of bronze, which required specialist knowledge and skill. Surpluses could also be used to improve agricultural productivity beyond that achieved at village levels, with major advantage when authority was placed in the hands of administrators who managed irrigation and other public works in the valleys of the Tigris, Euphrates, Nile or Indus.

Finally, the exchange of surplus produce for traders' and specialists' manufactured goods stimulated the growth of market centres which could develop not only as the concentration of non-agricultural specialists and traders but also as religious cult centres with a temple and temple priesthood. Where those in authority over temple complex, over commerce and over urban settlement came to overlap, we see the emergence of a state elite and state authority which would eventually be defined as kingship. As wealth was acquired and accumulated by a ruling elite, they could trade this for the labour required in major public works. Further competition and wealth transformed the urban power of Mesopotamian city states into a larger entity. At some stage, maintenance of law and order internally, and military protection against the power of other states, became part of the story.

More broadly, we see the beginnings of social class defined by specialised roles and social stratification defined by family wealth. And while we know less about the rural economy of Sumerian society than we do about excavated urban centres, the achievement of economies of scale would have aided increases in production and the creation of significant further surpluses.

The Urban Revolution saw the development of Sumerian towns with planned space to manage growing urban populations and monumental architecture including both temple complexes and residential complexes of those with authority over the town (whether defined as priests or kings). In Egypt, the concentration of surplus wealth is most visible in the royal and elite tombs which by ca. 2700 BC had developed into a massive pyramid construction.

But our image of urban life needs to retain awareness of the substantial agricultural region around a town, whose population provided the food for urban dwellers, were substantial purchasers of urban goods, and directly or indirectly created the surplus that kept the ruling groups of the city in wealth and authority.[4]

Such a model, as noted, does not answer the contested questions of cause and effect in the origins of what we label civilisation. But it was within this complex that the first writing in the world began, to fit the needs of the rapidly changing human world.

The Origins of Writing

Symbols before Writing

Writing means signifiers of sounds or words, not symbols showing the object represented. While writing is a phenomenon acquired by different societies at different times between 5,000 and less than 200 years ago, visual symbols have been part of human communication for tens if not hundreds of millennia. These may be scarification and tattoos on the human body, incisions on stone or distinctive markings of the creator on pottery. Visual representations we describe as 'art' in prehistoric societies may have filled a wide range of different social and personal functions. Upper Palaeolithic paintings of animals in the hidden interior of a European cave, or engraved animals on an open rock surface in Australia or the Sahara or the North American plains, may reflect a wide range of different meanings to the artist and the viewer. They may present an animal that is regularly hunted for food. They may show an animal that is a desired, but scarce, food resource. They may be a totemic affiliation. They may mark out or claim a territory. They can indicate identities in the stories that explain how the world works (what outsiders would call mythology) – as do figures that represent imaginary animal creatures and half-animal, half-people images. As with the San of southern Africa, they may represent the experience of a shamanic spirit possession.

Like animal images, so paintings, drawings and engravings of human figures may not be direct representations of what the viewer sees. Other images may reflect and record an actual event. And more abstract symbols, involving lines and shapes, can have a function or meaning that is clear to the creator and viewer but obscure to the modern observer. Some of this art may just be undertaken for pleasure, to ward off boredom or to show off a skill to a potential mate. The cross-hatched pattern made with an ochre crayon on a stone at Blombos cave in South Africa, dated to 73,000 years ago, shows us activity and ability but not its meaning.[5]

None of this is 'writing' in the sense that sounds are meant, though pictograms conveying the meaning of the object represented can anticipate the first writing of words, and, as in hieroglyphic Egyptian, they can supplement characters in writing to indicate the class of object implied by a word.

Writing can represent sounds but does not necessarily represent spoken language. The effort in creating a written form – a Sumerian clay tablet with cuneiform script or an Egyptian hieroglyphic inscription – imposes a formality on the substance and implies and ability by the reader to interpret the signs correctly. But throughout history, we can assume a difference between the language heard in everyday exchanges and the language that is heard when a written document is read out. Intonation differs, the words used differ, some formal words never used in conversation appear in written documents, and words in common exchange would be too informal to appear in written form. More formal contexts of writing may survive longer in the archaeological record than many informal documents for temporary use. And the use of symbols for recording business transactions as seen at the beginning of the phenomenon of writing represents something different from the words spoken to seal the deal. A reading out loud of numbers and measures is quite different from continuous narrative text. Early writing may represent commercial documents, labels or lists that were not an equivalent of a spoken conversation; even the dramatic text of a commemorative stela or inscription is far from the everyday speech of the administration; and early Sumerian texts which lacked the elements of adverbs or pronouns remind us of this contrast. Indeed, the first known writings, the clay tablets from the Uruk culture of the late 4th millennium, are so dominated by administrative documents recording products, quantities and the like that they cannot be said to represent speech as such.[6]

Before the first development of writing in Mesopotamia, the region had used impressions on clay – stamp seals – to mark ownership of pottery vessels. These are characteristic of the settlements in the pre-metal Neolithic Halaf period of Upper Mesopotamia (ca. 6100–5100 BC) and the Ubaid period which began in the alluvial plains of Lower Mesopotamia (ca. 6500–3800 BC), both periods named for sites where they were initially identified and defined.

Geometric patterns, rather than pictorial symbols, were the feature of these stamp seals and are therefore considered by some as ancestral to the first writing, although incised decoration in the region extends back for millennia earlier. A piece of stone engraved with a distinctive and individual pattern could be used repeatedly to stamp ownership on the clay seal of a pottery container or a coiled basket.

One well-studied collection of stamp seals from the Syrian village site of Tell Sabi Abyad, dated ca. 6000 BC, may have marked ownership of locally stored items, rather than goods to be traded.[7] Seventy-seven separate designs of seals were noted. When the stored items were opened for use, the clay seals were of course broken and discarded, to be recovered in archaeological work.

In the villages of the Ubaid culture, stamp seals were further developed with a very broad range of styles and symbols, though in use by only

The Origins of Writing

a minority of the community.[8] Their purposes were probably very varied, including sealing goods for sale, marking of ownership, recording an administrative or commercial agreement or as amulets and talismans. Naturalistic images are also found in seal impressions, which has led to the suggestion that these have a religious symbolism, perhaps as 'memory tools'.[9]

By the mid 4th millennium, stamp seals with symbols began to be replaced by more ambitious cylinder seals engraved on hard stones, which could be used repeatedly when rolled over a clay surface to create a distinct picture. Bearing often complex artistic images, such seals were at first purely visual, without lettering, and have provided a powerful tool by which the religions and societies of Mesopotamia can be interpreted. But the signs and symbols which served as aides-memoires were not yet writing.

Writing Numbers

From around 8500 BC, societies of Mesopotamia and adjacent regions from Palestine to Iran used clay objects – tokens – which themselves symbolised (but did not represent visually) the items of economic exchange, such as a quantity of grain or oil, a head of livestock or even an area of land.[10] These objects were of different geometric shapes and sometimes carried incision or punctate markings. By ca. 3500 BC, the usage had become complex enough to require spherical or ovoid clay 'envelopes' to hold them, with distinctive seals to close off the envelopes.

With increasing complexity of society came increasing complexity of economic relations. The rise of specialised craft, expanding trade, disparities of wealth and land ownership and the accumulation of possessions by a section of society meant that more involved processes of reckoning and accounting became essential in the urban culture of the ancient Near East, and the need to record these in a material form became highly important.

The writing of numbers, measurements and relevant calculations took precedence over the recording of words. Marking a quantity of items by a comparable sequence of notches on a bone or stick can be assumed far back into prehistory, but writing allowed a new approach to large numbers and measures.[11]

Larger numbers depend on an agreed base which would develop from a non-written method of calculating, using fingers and hands (or even toes). Our own decimal system of units–tens–hundreds differs from those in ancient Mesopotamia, which included multiples of tens but primarily

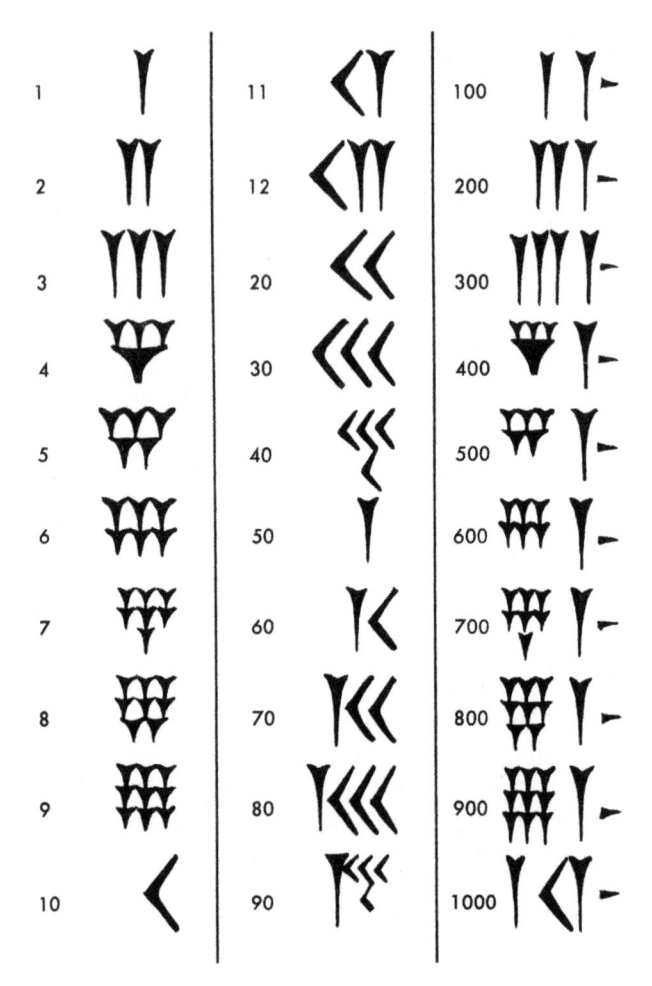

FIGURE 15 Sumerian numerals in cuneiform (photo credit: Science History Images/Alamy Stock Photo).

adopted a sexagesimal system with sixty as a base. This may seem strange to us, but note the practical point that sixty has eleven divisors, whereas ten has just one, two and five. (If the human hand had been devised to help develop mathematics, it would not have had five fingers.) We have inherited and retain aspects of the sexagesimal system today, with our globe divided into 360 degrees, our hours into 60 minutes, and our minutes into 60 seconds.

From the earlier use of clay tokens representing numbers and commodities, Sumerians of Mesopotamia developed simple symbols created using the end of a stylus.[12]

The Origins of Writing

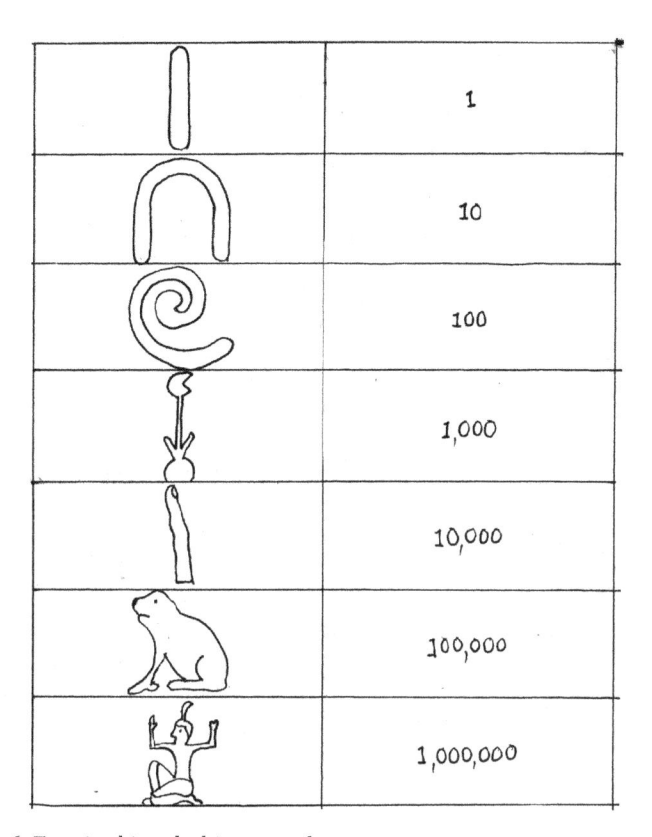

│	1
∩	10
⌒	100
⚘	1,000
↑	10,000
𓆏	100,000
𓁨	1,000,000

FIGURE 16 Egyptian hieroglyphic numerals.

Many of the earliest Sumerian clay tablets were accounting documents recording quantities of trade items and incorporated symbols to represent the relevant items. One tablet includes two exercises to calculate detailed field measurements.[13] The early cuneiform script subsequently developed by the Sumerians presented as numbers a combination of vertical impressions for numbers one to ten, and a variety of strokes for multiples of ten or sixty (Figure 15).

Egyptian numbers were based on the decimal system, and hieroglyphic Egyptian used a single stroke for a unit and symbols for higher numbers (Figure 16). Egyptian numerical symbols were found alongside short text signs in the late predynastic tomb U-j at Abydos, and with early text writing in other late predynastic and Early Dynastic contexts. Egypt maintained through the history of pharaonic Egypt the same symbols for numerals seen at the dawn of Nile Valley civilisation.[14] The only difference would be that the use of the term for a million also came to mean the value of 'a multitude'.

Pictograms, Ideograms and First Writing

Systems of writing sounds and words were independently developed in several locations, but the oldest writing is from the ancient Near East: Mesopotamia and Egypt, with evidence for writing from both by around 3100 BC. Tracing the development of these writing systems and their impacts and contributions to their societies shows the importance of the innovation of representing sounds and meanings by symbols whose meanings could be learned and readily interpreted.

Single images had long served to convey a message concerning the object shown. But the idea of an image to represent sounds rather than to depict objects is revolutionary; the innovation of writing was a change of thinking and a change of direction from the representational.[15] The beginning of writing responded to the need to convey a more complex message than 'this is mine', or the use of a stylised image of an ox to mean 'ox'.

The first real writing can be defined as the use of written symbols to signify something beyond what was directly shown, and in a code in which the wider meanings were readily understood by others. Describing the first writing as 'pictographic' implies this: an abstract concept depicted by a recognisable image.

The earliest pictographic writing is associated with the Sumerian languages of southern Mesopotamia and the culture named after the type site of Uruk, at the time the largest settlement anywhere in the world (ca. 3800 to 3100 BC).[16]

Early Sumerian society was dominated by separate city states in the alluvial plain of the lower Tigris and Euphrates, where the agricultural wealth of the region supported specialists, temples and urban elites. The advantages of transport on the river system were a major contributor to both trade and administration and to the creation and accumulation of wealth.[17] Whether we call them towns or cities, urban settlements grew in size; Uruk itself would extend to 5.5 km^2 (2.2 sq ml), with tens of thousands of residents. The cultural region extended into Susiana in modern Iran and north into the broader area of Upper Mesopotamia.

The largest collection of archaic written texts available for study was excavated from the mound around the Eanna temple in Uruk, but frustratingly it is difficult to assign an exact date to their creation because their context was that of rubbish dumped in antiquity, not the actual setting of their original creation. Some 5,000 texts have been recovered from this context, contributing to a corpus of over 5,800 texts for study and analysis.[18] A date around 3200 BC, during the so-called Uruk IV phase, seems likely for the first stage of pictographic writing.

Archaic pictographic writing (sometimes call proto-cuneiform) emerged in the middle of the Uruk period, in the later 4th millennium BC. A priority in early Sumerian writing was the recording of transactions relating to materials, or property or labour – topics which required a combination of words and numerals. The early pictograms do include some characters which picture things they represent.[19] Signs for livestock include associated signs to indicate categories of sheep or goats. But they may also be indicative of the meaning: thus, a sign bearing resemblance to a foot was used to indicate going or standing.

Each symbol used in Sumerian writing, both in archaic forms and in more developed cuneiform, typically represented a syllable by each symbol. Ideograms served to signify something that could not readily be represented pictographically. As writing developed, a sign would be used as a 'rebus' or homophony, where the syllabic sound indicated by one meaning is used in another meaning. For example, the Sumerian word for fish was *ku* and was represented by a stylised fish. But that same symbol could be used to be read as the syllable *ku* in other contexts.

Gradually, the Uruk Sumerian pictographic and ideographic symbols for words and sounds became simplified or transformed to make their writing easier. The successive period of Sumerian development, Uruk III (also called Jemdet Nasr), dated 3100–2900 BC, saw development of writing with more abstraction in contents and fewer curves in execution. Drawing individual lines on clay was a laborious process and limited the images that could be clearly shown, so impressing a stylus on the surface allowed a broad range of symbols This flexibility marked a move away from the use of images reflecting a real-world object, towards abstract shapes. Such a shift marked a stage towards the long-lasting writing style we call cuneiform. Study of the Eanna archive was able to distinguish these two separate stages in the development of writing.

There have been different estimates of the number of signs in the archaic pictographic scripts. In addition to 60 signs representing numerals, some 1,900 symbols have been suggested. But when variants and reverse orientations are removed, a number of around 900 separate signs may be realistic.[20]

The script changed gradually during the succeeding Early Dynastic period of Sumerian civilisation (ca. 2900–2350 BC), the era best known for the riches of the Royal Cemetery of Ur during which the area of southern

Mesopotamia was at its most urbanised. During the Early Dynastic, the forms of both numbers and text were simplified, as shown in finds at the southern Mesopotamian sites of Fara (ancient Shuruppak) and Tell Abu Salabih (Salabikh), 160 km and 220 km (100 and 135 miles) south-east of modern Baghdad.[21] The thousand tablets from Fara are dominated by administrative texts but include lexical and school training tablets. By the middle of the 3rd millennium BC, true cuneiform writing had emerged.

The Development of Cuneiform

It was a laborious task in archaic Sumerian writing to draw and impress multiple pictographic and ideographic symbols in clay in such a way that the symbols of a document were unambiguous. The shift to writing entirely with impressions made with a stylus eased the whole process, and thus cuneiform was developed (Figure 17).

Cuneiform – meaning wedge-shaped – was not a language but a script, a means of creating writing. It proved so convenient and adaptable that it was used to write in a wide variety of western Asiatic languages between its emergence in fully developed form by the mid 3rd millennium BC and its final uses in the 1st century AD. A stylus was impressed into a clay surface in a combination of horizontal, vertical and angled impressions to create a single symbol. Even if the stylus was not itself of wedge profile, it was applied to give the wedge-shaped impression.[22]

By around 2500 BC, the number of signs seems to have been around 600 – significantly fewer than would be used in Chinese script. The majority of these symbols represented syllables, but some served as determinatives (indicating the class of object referred to) and some represented whole words (logograms). Once fired, the clay (most commonly clay tablets) held the message in permanent and unalterable form. The characters used in developed cuneiform writing were no longer recognisable in their derivation from pictographic symbols; and (unlike Egyptian, Greek or Roman lettering) curves had disappeared from the written form.

While cuneiform began in order to represent the sounds of words in Sumerian (a language with no known links), it needed to be adapted to represent the very different language of Akkadian, one of the many languages of the Semitic family that would be found across western Asia.[23]

The cuneiform that developed by Early Dynastic III Sumer is reflected in a richer archaeological record in the period of Akkadian domination of Lower Mesopotamia (2350–2200 BC), beginning with the rule of Sargon, and in the return to the rule of the region by Sumerians under the Third

The Origins of Writing

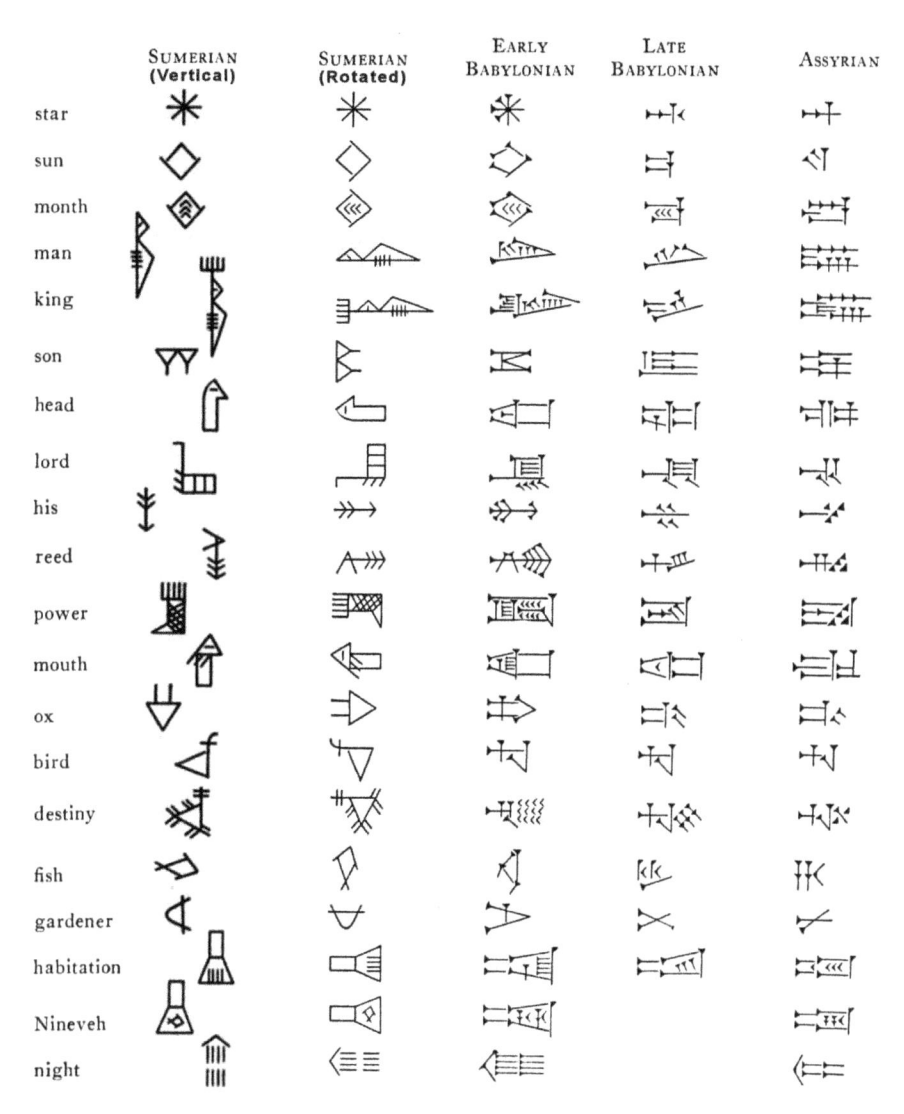

	Sumerian (Vertical)	Sumerian (Rotated)	Early Babylonian	Late Babylonian	Assyrian
star					
sun					
month					
man					
king					
son					
head					
lord					
his					
reed					
power					
mouth					
ox					
bird					
destiny					
fish					
gardener					
habitation					
Nineveh					
night					

FIGURE 17 The development of cuneiform writing (W. A. Mason, Public domain, via Wikimedia Commons).

Dynasty of Ur (2112 to 2004 BC). From that later period, more than 90,000 cuneiform documents are available for study, though records relating to the agricultural sector still dominate. In total, 30,000 tablets were recovered from the site of Nippur, with most from the 3rd millennium BC, representing a 'scribal quarter' in the Third Dynasty of Ur period.[24] By this era, documents have added value to history by including the year name (identified by a prominent event).

The succeeding Babylonian and Assyrian empires which ruled Mesopotamia used cuneiform writing to present texts in their own Semitic languages. Major archives of their texts serve to demonstrate the broad range of uses to which writing was put but also include copies of earlier Sumerian texts of value to us in tracing early literature, political authority and religious belief.

Hieroglyphic and Hieratic

Mesopotamia and Egypt developed urban civilisations with ruling elites and state structures (though in differing patterns) at the same time, the late 4th millennium. The first evidence of writing is found in each society also at this time, but in different writing styles and with different material evidence to demonstrate its uses. Proto-writing may be indicated by signs which appear on Egyptian cylinder seals and on clay sealings from the period 3400–3300 BC. Over 500 inscriptions have been published from the period before the beginnings of the First Dynasty at ca. 3100 BC.[25]

There are different estimates of dates for the predynastic and protodynastic cultures of Egypt's 4th millennium. One convention describes the period, also known as Naqada III, of rulers grouped in a Dynasty 0 (3300/3200–3100) of kings (Iry-Hor, Ka, and 'Scorpion') before the First and Second Dynasty. These (together the 'Early Dynastic') dated from ca. 3100 to 2700 BC. The Third Dynasty from 2700 BC, beginning with Pharaoh Djoser, marks the pyramid age of the Old Kingdom, which lasted for half a millennium until 2200 BC.[26]

The valleys of the Nile and the floodplain of the Tigris and Euphrates were in trading relationships across the 1,500 km (ca. 930 miles) distance that separated them: a distance of at least 1,800 km (ca. 1100 miles) by 4th-millennium trade routes through the Levant and the Upper Euphrates, where local traders may have been the active intermediaries. Alongside materials that were traded along such routes, innovations and visual themes from Mesopotamia are seen in the period that marks the beginning of writing in Egypt, and cylinder seals of Mesopotamian type are found in Egypt even earlier. A terracotta cylinder with the name in hieroglyphic of a First Dynasty queen, Mer Neit, was found at Uruk.

Therefore, it is reasonable to suggest that the 'idea of writing' was passed between them by travelling merchants, especially as it was commerce that provided the incentive for the first steps to record transactions in

The Origins of Writing

Mesopotamia. Similarly, the development of the Minoan scripts of Crete in the 2nd millennium (Linear A and Linear B) reflects the idea of writing without copying from other scripts.

Early Mesopotamian writing was by impressions on clay, which was then baked and so made permanent. Because clay is a common resource, writing on such an available material had no cost other than time. By contrast, if Egyptians wrote commercial or administrative documents on papyrus or linen, they were using up a manufactured material with greater cost and working on a material which disappears much more readily from the archaeological record.[27] That may explain why we see so many early documents from Mesopotamia and so few from Egypt.

The use of papyrus is affirmed. Blank rolls of papyrus have been found at Saqqara from the tomb of Hemaka in the Early Dynastic period, and a sealed papyrus scroll is a hieroglyph from the period too.[28] An important archive of papyrus administrative records was found in 2013 at Wadi el-Jarf on the Red Sea, dating from the reign of Khufu (Cheops) and relating to material for constructing the Great Pyramid of Giza; it is evidence of the routine use of papyrus for administrative purposes. Given the scarcity of contexts in which papyrus could survive, the first writing in Egypt might therefore be even earlier than that in Mesopotamia.[29]

Another relevant aspect of evidence is that most of our archaeological materials for late 4th-millennium BC Sumer comes from settlements, which provided the documents of everyday economic and social relations. By contrast, the majority of material from Egypt comes from burial sites, which were typically located just beyond the cultivable land of the Nile.[30] Writing for the dead is a different context from writing for the living; absence of evidence is not evidence of absence, and early economic documents on temporary materials from Egyptian town and village are missing from the archaeological record. Where later predynastic and Early Dynastic sites have been excavated in the Nile Delta (Lower Egypt), written material has not survived.

Egyptian writing developed quite differently from that of Mesopotamia, where pictographic images were gradually transformed into cuneiform symbols whose origins were unrecognisable. Egyptian formal writing – which we call hieroglyphic – began with the use of recognisable pictorial images for letters, syllables and determinatives. While the very first hieroglyphic signs differ from the developed pharaonic hieroglyphic, by the Early Dynastic era the signs include forms which remained for over three millennia (with a final specialised use in the 4th century AD). Over time, there was only a modest number of changes (and temporary introductions) to the signs in use, despite the evolution of the Egyptian language. Meanwhile, as well as hieroglyphic, a more cursive form was used, easier to write with a reed pen, which we call hieratic. Hieratic began early alongside hieroglyphic and was the commonest form of daily practical writing in pharaonic Egyptian society until the mid-1st millennium BC.

The firmest evidence for the first writing with hieroglyphic signs comes from the excavations at Abydos of cemeteries U, B and Umm al-Qa'ab, 11 km (7 miles) west of the Nile and 400 km (250 miles) south of the later pyramids of Giza. Abydos is also the location of burials of the first pharaohs.

Here, an impressive elite tomb named U-j, from the latest predynastic period (Naqada III, with a suggested date as early as 3300–3250 BC), was discovered in 1988, with details published a decade later.[31] It comprised twelve subterranean chambers extending over 125 m². Familiarity with writing was shown by the presence of hieroglyphic signs on some 200 bone and ivory labels (tags) to be fastened to commodities, and there were also simple painted signs on over 100 pottery vessels for oil (Figure 18). The tags carry between one and five hieroglyphic signs, the pots just one to two signs.[32] The writing represented names of people (given the context presumably including people of influence), place names (including places interpreted as towns in the Lower Egyptian delta) and numbers of items. The similarities of the ivory labels suggest they were applied by a single administrative centre, in the area of the goods' receipt – not labelled ahead of trade or despatch.

We can therefore see the first Egyptian writing as a recognisable shorthand representing sounds (as in names of objects, people or places) but not that of continuous speech, which we see half a millennium after the first hieroglyphic signs on tags.[33]

A feature of hieroglyphic and hieratic texts is their signs representing single, double and triple consonants only, without vowels indicated. The two dozen alphabetical signs were given no special priority. A group of signs might therefore represent a range of very differently pronounced words – the determinative sign (to indicate the class of object meant by a word) could guide the reader on which was meant (and indicated the end of a single word), but the overall reading of a sentence would depend on the understanding and interpretation of the reader.

The Abydos ivory labels included signs for two-consonant sounds. Some signs representing three consonants had developed by the First Dynasty, and determinatives were well established by the middle of that period. The first complete sentence is only known from the Second Dynasty, and it would wait until the Old Kingdom from ca. 2700 BC before it was normal for texts appear as representations of sentences in speech, with verb

The Origins of Writing

FIGURE 18 Umm el-Qa'ab tags (Deutsches Archäologisches Institut).

conjugations and prepositional forms. While the uses of writing thus changed and grew, the number of signs was reduced from up to 1,000 seen in early texts to about 500 used in 'classical' Middle Egyptian.

Already by the reign of Sekhen-Ka of Dynasty 0, ink inscriptions on pottery identify 'accounts of Upper Egypt' or 'deliveries from Lower Egypt'.[34] Hieroglyphic signs are present as part of the dramatic carved reliefs which mark the beginnings of pharaonic rule (Dynasty 0 and the First Dynasty) on stone palettes and mace-heads, including spelling out the names of pharaohs. Visual representations are typically accompanied by names in signs which cannot always be interpreted, such as those at the top of the Hunters' Palette.

Royal and elite burials are attested at Abydos, at Sakkara (where there are cenotaphs to the pharaohs of the first two dynasties – argued by some to be their real tombs rather than those at Abydos) and also at Hierakonpolis, all in Upper Egypt. While the impact of these ceremonial items was visual, the presence of names of people or places implies a relevant level of literacy. The system was still in a developmental stage: the Horus name of Narmer (either the final Dynasty 0 ruler or the first ruler of the First Dynasty) is shown in fifteen different sets of signs in the texts.[35]

In addition to inked names on pottery and tags (labels) to mark commodities and sales, Early Dynastic texts include steles on tombs giving names and titles, rock inscriptions and some carving in stone elements on brick buildings. An unusual document from a First Dynasty burial site is a list of the names of the first kings.

From the beginning of the First Dynasty, we see a feature of hieroglyphic writing that would become distinctive: the organisation of the position of signs to make an evenly composed and balanced register. Early examples have been studied from museum collections and formal excavations.[36] During the first 500 years of Egyptian writing, until the reign of Djoser (builder of the Step Pyramid) in 2700 BC, the texts we have are typically brief, with 4,000 inscriptions known containing only 20,000 characters, representing a summary of ownership, a transaction or a caption for an image rather than a continuous text.[37]

It is unsurprising that the complexity of painting or carving representational images in hieroglyphic script invited the quicker hieratic form of writing in daily use. This seems to be the form used in texts on numerous Second Dynasty stone vessels beneath (and therefore earlier than) Djoser's Step Pyramid at Sakkara, noting people, places and accounting information.[38] These also mention ceremonies in which the vessels (or presumably their contents) were used. Such scripted ink handwriting could be called proto-hieratic, and by the Old Kingdom hieratic writing became widely used.

The Idea of Writing

We have noted the common suggestion that 'the idea of writing' spread between Egypt and Lower Mesopotamia. With the trade in material and manufactures between the two regions, and the focus of initial writing on commercial activity, it is unsurprising that the idea of a shorthand simple action to create shared meaning would seem appealing. Egyptian hieroglyphic and hieratic writing spread across the area controlled by emerging Nile Valley rulers and also into the Levant, reflecting active trade networks. Cuneiform would be applied to represent different languages, including that of the Hittites of Anatolia. Soon after the first writing in Mesopotamia, a different form of writing was to be found from ca. 3100 BC in Iran – the Proto-Elamite sites of Susa, Tel-e Malyan (Anshan) and elsewhere. This too used impressions on clay tablets. The Proto-Elamite script, whose numerical notation echoes that of Mesopotamia, seems to have developed rapidly into a full system of writing once its value in administrative roles, as shown in neighbouring Sumer, was clearly recognised. By 2300, a new writing form, Linear Elamite, had developed, not derived from the cuneiform model, with symbols possibly all phonographic, representing only sounds.[39]

The 3rd-millennium trade links between Mesopotamia and the Harappan civilisation of the Indus Valley support the view that the idea of writing spread east to influence its local development. The Indus Valley script itself is not linked to cuneiform, and efforts continue to achieve its decipherment.

In Crete, with strong trade-route connections to Egypt, the distinct writing styles of Linear A (still not deciphered) and Linear B reflected the Egyptian idea of written forms without the use or influence of the Egyptian writing style.

While hieroglyphic included alphabetical signs alongside others, it was never reduced to make them the basis of writing. Alphabetical writing is first seen in the Proto-Canaanite inscriptions seen in Sinai (in the early 2nd millennium BC) and in subsequent developments in the first South Arabian script and the Phoenician alphabetic script which was transformed into Greek and Roman lettering.[40]

An alphabetic (though strictly consonantal) script was identified at Ugarit (Ras Shamra) on the Syrian coast and dated from the 15th century BC onwards. Initially created with cuneiform impressions, its thirty signs were designed primarily to represent the local vernacular languages. The same place and time had documents with at least eight languages in five scripts.[41]

Other Independent Origins

There are different definitions of what characterises a true system of writing, and many even relatively complex societies have operated without the need for a written form, though adopting writing may itself act as a catalyst for further social complexity.[42] Different histories show the cultural importance assigned to writing once the idea of its power had spread. Pride in a community language can be signified by pride in a community script. The ancestry of scripts such as Hebrew and Arabic can be tracked back through Middle East sources, but there are other scripts which copied the idea of writing but not an existing form.

The independently developed markings used in the eastern Pacific territory of Rapa Nui (Easter Island), *rongorongo*, are considered by many an independent invention of writing. The writing developed for the Native American Cherokee language in 1821, with eighty-five characters representing syllables, was a new script within a society which already had another available. In the 1830s, non-literate residents of Liberia in West Africa created the syllabic Vai script, with about 200 signs, which has continued in some use until the present day.[43] The short history of this is an indicator of how some of the earliest developments of new scripts may have taken place.

Writing developed – was invented – in China without any influence of the idea from elsewhere. Chinese is considered the oldest script to continue in use today, presenting a challenge to learners in its traditional form with its thousands of characters. It is first visible in the archaeological record from the Shang Dynasty period, when we have inscriptions from the 13th century BC on oracle bones or tortoise shells used for divination (Figure 19). The inscriptions begin with basic information on the time, place or personnel of the divination, then the statement being tested by the divination. When the bone or shell was heated, the resultant cracks provided the answer to the issue raised in the text.[44] Only a little later, the script is found on Shang bronze vessels (carrying the name of the maker or a dedicatee commemorated by the vessel).

Since the script seen on the oracle bones appears well developed, it is reasonable to assume its initial use somewhat earlier, for different and arguably secular purposes, on organic materials which have not survived.[45] Signs and symbols marked on pottery appear in earlier contexts (the Late Neolithic of the 3rd millennium BC): in form, they do not appear ancestral to the oracle bone script.[46] Nevertheless they may have served to convey information between different communities in a trading

The Origins of Writing

FIGURE 19 Oracle bone (National Museum of China, CC0, via Wikimedia Commons).

relationship. This is not writing in the sense of conveying sounds and spoken language but can be compared with the initial uses of written symbols in the Near East, where they do appear to have developed into a full written script.

Early signs in this Chinese writing can be described as initially pictographic, but some signs developed to present syllabic sounds. About 4,500 characters have been distinguished on oracle bones, with some different symbols carrying the same meaning.[47] The script developed with transition of pictographic signs away from realistic imagery and signs, predominantly representing syllables, or syllables with their meaning qualified, without clear pictographic origins. By the 1st century AD, a dictionary could list 9,353 different characters.[48]

There was no inevitability that an advanced society with a dominant state structure would develop writing. The Inca of Peru used *quipu* – knotted strings – as a method of recording information. Aside from the Old World, the other major independent invention of writing was that of Mesoamerica. Different written forms occurred widely across time and space in the Mesoamerican region, and Maya glyphs continued in use until the period of European contact.[49]

The origins of writing in Mesoamerica are still not clear, but it appears to have occurred before the presence of centralised state societies. The Olmec people occupied this region from ca. 1500 to 400 BC, the Zapotec thriving from ca. 700 BC, and finds of writing have been identified from both Olmec and Zapotec contexts.

A block of serpentine rock from the Olmec site of Cascajel with a possible date of ca. 900 BC carries twenty-eight different signs whose arrangement suggests a formalised word order.[50] Somewhat similar characters are inscribed on stone from Tlaltenco and Humboldt. If inscriptions had been made then or earlier on wood, they would probably have decayed. Further examples of glyphs from Olmec sites have been cited as evidence of developing writing systems.

The Olmec and Zapotec scripts do not appear ancestral to the other and later known scripts of the region. It seems reasonable to suggest the Maya adopted the concept of writing from earlier scripts in southern Mexico, just as the *idea* of writing may have been transmitted between Mesopotamia and Egypt but adopting different forms.

Classic Period Maya writing (between 250 and 900 AD) was used for a wide range of functions, from the description of events to the details of a calendar. Mayan writing was found on monumental architecture (stone and stucco), jade and ceramic objects and carved bones and was written on cloth, wood, skin and bark paper surfaces, with more unpredictable survival.

Late Classic Maya script has over 300 signs. Characters represent either syllables or whole words for objects; some symbols could have both uses, and while most Classic Maya texts represent the language of the Maya priesthood who acted also as scribes, there are variant Mayan languages represented in the same script. Debates on the origins and sequence of Mesoamerican writing will continue.

The Early Impacts and Roles of Writing

In the previous chapters, we noted two stages of innovation. Human use of wildfire was followed much later by the ability to create and curate fire at will. The role of the domesticated horse saw a major change when it was attached to wheeled vehicles. Similarly, we can suggest stages in the development of writing in the ancient Near East as a facilitator and catalyst for many social activities. The first uses of the symbolic representation of sounds in both the Nile Valley and Lower Mesopotamia were at the end of the 4th millennium BC. A wider range of applications is attested in the more complex societies that developed over the next six centuries, before the last stage of Early Dynastic Sumer (ca. 2500 BC) and the beginnings of Old Kingdom Egypt (2700 BC). The majority of scripts in use in modern times trace their ancestry to the invention of writing systems in the Middle East at the end of the 4th millennium BC, and it is their impacts which we explore.

In the longer cycle of ancient Near Eastern civilisations and later, the value and applications of writing were many and clear. What is important is to consider the early transitions that came about from the invention and use of the innovation: to assess not what roles writing eventually came to serve, but what were its earliest impacts and functions, and its advantages to a society in a period of dramatic change.

While later myths sought to explain the divine or regal origins of writing, in the first stage we can note that 'writing had little or nothing to do with oral discourse; it was an administrative technology that extended the denotative versatility of language into realms that had previously been served only by the drastically more limited devices of tokens and seals'.[51]

Ownership, Identity and Trade

A pictorial symbol on a tag attached to an object, or stamped on the seal of a container, only indicates the owner or source to a small number of individuals who know that 'the rosette on an item in that storage area means it belongs to X'. Written signs whose meanings are *sounds* convey to anyone the name on a label – and that label once affixed retains its ability to convey that meaning at any stage of a commercial activity from its first packaging to its final delivery. No insider knowledge is required; just a level of literacy adequate to read the written or inscribed signs as sounds which reflect the name (whether of person or place).

The nature of a product and the quantity involved can also be part of a label: 'X amount of barley supplied by Y', integrating numerical systems and names with words whose written form may initially have resembled the substance indicated, but which would soon be adapted to a more easily written or impressed letter in a script understood by all parties.

It is just one stage further from an exchange of goods to recording an agreement to trade – the names of both parties alongside produce and quantity. Such a contract, summarised in text impressed on a clay tablet then made permanent by firing, acts as a formal confirmation beyond that of verbal exchange. It maintains the record of that commercial agreement if necessary across a season when produce was stored for future use. Private documents from Sumer record the numbers of items in a transaction; the administrative documents of institutions required a more complex method of record keeping.[52] The Egyptian ivory tags from Abydos seem to represent receipt and use, in this case in an elite tomb, but we can assume similar labelling of goods at earlier stages of the trade transactions.

Land, Rent and Building

Ownership and agreement relating to the use, lease or sale of land could not be attached physically as with goods, but an unambiguous record of a transaction could be made in permanent form in writing. The fired clay tablets of Mesopotamia for such agreements have survived, while the equivalent early Egyptian documents do not.

A combination of signs for numerals, measures and text provides records of calculating costs, space and value in Sumerian cuneiform tablets, alongside visual presentations. The earliest Sumerian documents of the Uruk period are dominated by accounting records, but with writing not only the numbers of items could be recorded but also associated abstractions: time, place, personnel involved and relevant administrative action undertaken.[53] Likewise, the archaic texts of the Uruk III (Jemdet Nasr) period record transactions in land, the storage and distribution of harvested grain, information on other commodities and also the names of individuals involved in these transactions.[54]

More dramatic than the clay tablets are the rare Sumerian inscribed stones recording, presumably *in situ*, the transfer of land.[55] These appear to prefigure the boundary stones known from later Babylonian times. That of a priest Ušumgal and his daughter records a transaction involving three fields, three houses and some livestock (Figure 20).[56]

The Early Impacts and Roles of Writing

FIGURE 20 Stone stele of Ušumgal, ca. 2750 BC, marking real-estate transfer (CC0, Metropolitan Museum of Art).

The ability to make, record and transmit complex calculations is relevant not just to land and property, but also to the work of surveyors, engineers, architects and builders. A long evolution of building developed into the stepped ziggurats of Ur and the pyramids of Egypt's Old Kingdom, requiring ever greater complexity in planning their design and construction and conveying these plans across a large team. The first pharaoh of the Old Kingdom, Djoser (ca. 2700 BC), had a step pyramid built at Sakkara some

330,000 m³ in volume on a 120-metre base, but no record of the architect/ engineers' plans and calculations exists from that time. A much later papyrus from the 12th century BC now in the Turin Egyptian Museum has text and images for the tomb of Pharaoh Ramesses IV, showing what we are missing from the much more ambitious monuments of the pyramid age.

Administration and Authority

Arguments continue about the nature of authority and the state at the time of the earliest writing in Lower Mesopotamia; and the same period in Egypt has an emerging pattern of rulers possessing (or claiming) power over increasing territory. In both cases, the role and contribution of centralised authority included water management, in which the benefits to individual farmers came from the ceding of authority over a large area of cultivable land to create canals and irrigation.[57]

In Egypt, management of water resources also involved measurement of the annual Nile flood, when the flow in the Blue Nile from the Ethiopian highlands reached the Nile Valley. Recording the time and height of the annual flood to predict and plan the annual cycle of irrigation and agriculture was a major responsibility of the nascent state.

In the archaic texts of the Uruk period, the frequency of words provides a fascinating picture of the society and its priority for record keeping.[58] Apart from names of deities, the most common words appear to be barley, distribution, overseer, female slave, delivery and what has been translated as 'accountant' – a rather grandiose name for what in other contexts we might call a bookkeeper.

The ambiguous (and evolving) term EN initially appears twice as often as any other word. It represents the most senior person in the temple system. EN is often translated by the neutral term 'chief administrator', which avoids the question of how much civil (or religious) authority he held. With the growing power of those who controlled the temple, its wealth and the society and resources it managed, a less democratic sounding term might be appropriate.

Some scholars, however, have suggested that the early writing in Mesopotamia came at a time of egalitarian social organisation, with the EN holding an agreed executive and administrative role, and city kingship developing only later.[59] This argument has been advanced by Graeber and Wengrow in their innovative book *The Dawn of Everything*. They note that there is no necessary correlation between size of settlement and the evolution of social complexity, given the large prehistoric communities that

The Early Impacts and Roles of Writing

existed without becoming states. They suggest there is no evidence for a monarchy in Mesopotamia until ca. 2800 BC, supporting the idea of autonomous self-governing units practising primitive democracy.[60]

A similar view had been advanced by German scholar of ancient Mesopotamia Petr Charvát, who argued that the urban societies of the Uruk culture and the succeeding Early Dynastic I and II were relatively egalitarian communities with administrations managing 'corporate undertakings for corporate use' (including waste management) and with kinship groups and kinship alliances the dominant features even within the growing urban settlements. Production surplus was sufficient to allay the costs of centralised administrative functions. By this model, the hierarchical structure of the state emerged only in the Early Dynastic III-period ruling elite, with the role signified by the term LUGAL or king now more important than the administrative EN.[61] The king could now influence temple appointments (as would be found in Egypt): a different model from the idea of temple priesthoods gradually gaining secular power.

If the image of the powerful state is a trope of the early 20th century and the idea of the democratic community is a motif of the early 21st century, there may be a compromise position in interpreting state formation. Arguments will remain.

Whether because of the authority of a powerful state apparatus or collaboration in more egalitarian societies, intergroup violence appears to have reduced with the beginnings of urban civilisation in the Middle East for a period, from ca. 3200 BC, before rising again from ca. 1500 BC.[62]

The link of EN to the central temple cults of emerging city states is reflected in the many texts which involved the temple in economic transactions, as the power of the temple and its associated personnel grew. Complexity and numbers of such economic transactions needed a system of record keeping that was unambiguous, and communication that could cut across time and place. Time was relevant: an agreement made could require payment or activity at another time, and a lease could be for a given period. Writing on baked clay tablets facilitated and assured this transaction. We see the growth of writing; we see the growth of bureaucracy; we see the importance of time-resistant record keeping and we see the emphasis on bookkeeping (or accounting) (Figure 21).

The growth of relative power is shown in the variation of tomb sizes and contents in final predynastic and Early Dynastic Egypt, though our direct evidence of writing from early settlements is limited. The centralised power in Egypt's administration required revenue, and this implied taxation. Tax

FIGURE 21 Sumerian clay tablet with administrative account of barley distribution (CC0, Metropolitan Museum of Art).

marks are already found on Egyptian vessels of Dynasty 0 and the First Dynasty.

In the earlier periods of writing, we do not see formal political records other than public affirmations for purposes of asserting authority in public declarations. We must assess the political reality through our interpretation of documents with other purposes. One example is the stele from the Egyptian Sixth Dynasty pharaoh Pepi I, which declared that the mortuary chapel dedicated to his late mother would be exempt from tax.[63] This indicates that taxing the revenue of religious centres was the norm.

A dramatic find in 2013 provided evidence of Egypt's Old Kingdom administration. A group of caves used as storage galleries inland of Wadi el-Jarf, a harbour on the Red Sea coast of Egypt, has given us a unique image of writing as an everyday activity during the Old Kingdom, when the Great Pyramid of 26th-century BC pharaoh Khufu of the Fourth Dynasty was being built at Giza.[64] Archaeologists recovered rolls and fragments of papyri containing writing (some in hieroglyphic and most in hieratic) relating to the activities of sailors and workers at the settlement. The location had been established to import valuable minerals from Sinai. Many of the papyrus documents detail the commodities supplied to these workers. Others are logbooks which record daily activities and those of an official called Merer, a man who had documented earlier work where he was responsible for

The Early Impacts and Roles of Writing

shipping limestone blocks from the Tura quarries north along the Nile for the Khufu construction.[65] A feel for the daily operation, management and supply of cohorts of workers is revealed by this remarkable discovery.

> First day the director of 6 Idjer[u] casts off for Heliopolis in a transport boat-iuat to bring us food from Heliopolis while the Elite is in Tura. Day 2: Inspector Merer spends the day with his phyle hauling stones in Tura North; spends the night at Tura North. Day 3: Inspector Merer casts off from Tura North, sails towards Akhet-Khufu loaded with stone. [Day 4 . . .] the director of 6 [Idjer]u [comes back] from Heliopolis with 40 sacks-khar and a large measure-heqat of bread-beset while the Elite hauls stones in Tura North. Day 5: Inspector Merer spends the day with his phyle loading stones onto the boats-hau of the Elite in Tura North, spends the night at Tura. Day 6: Inspector Merer sets sail with a boat of the naval section of Ta-ur, going downriver towards Akhet-Khufu. Spends the night at Ro-She Khufu. Day 7: sets sail in the morning towards Akhet-Khufu, sails towing towards Tura North, spends the night at [. . .] Day 8: sets sail from Ro-She Khufu, sails towards Tura North.[66]

The increasing complexity of Mesopotamian and Egyptian society involved a growing class of civil, military and religious officials. A powerful state required an effective administration of coordinated managers, officers, priests and scribes. A small city state might manage communication between rulers and senior officials by word of mouth, but a larger entity could only hold together if there was effective regular and frequent communication between the centre and the periphery. The rulers of Egypt from the First Dynasty claimed control over the whole of Upper and Lower Egypt. Earlier, the city state of Uruk had extended its sphere of influence northwards to other city states. Written information could remove the ambiguities, uncertainties and unreliability of the courier's memory: an instruction or a report on papyrus, linen or fired clay tablet was unambiguous.

Education and Lexical Reference

Although 'scribe' is not a common word in the earliest, archaic Mesopotamian texts, by the 26th century BC we start to find numerous references to scribes in the literature.[67] These convey the image of an exclusive craft in which a small number of privileged specialists were instructed. It is convincing to think of the creators of our very numerous inscribed clay tablets as part of a professionally trained trade, providing services to secular and temple administrations, merchants and citizens.

For normal communication in pharaonic Egypt, scribes had the convenience of hieratic script, although hieroglyphic symbols could be drawn in ink. Finely carved hieroglyphs in relief on stone objects were clearly the work of a different kind of specialist craftsperson, not just a scribe.

Texts created by scribes were of no value unless they could be read by others, not least the officials spread through the administration, even if these used scribes to inscribe a response. Commercial agreements needed shared recognition of their contents. Literacy – the ability to read cuneiform or hieroglyphic or hieratic documents – must therefore have been far wider than the professional scribes. A wide mandarinate was required by ancient states, even at stages of their early development, and the EN (whether an administrator or within a ruling elite of Mesopotamia) would need to read what scribes had written. In order that writing could achieve its purpose, scribes and officials alike would need to recognise the same group of signs conveying the same meaning – whether a name of a place or a personal name or an object or an abstraction.

It is unsurprising that a focus on the ability to read and write was the core of ancient education. We have some remarkable evidence of the documents used in teaching literacy in what are sometimes described as scribal schools but which, by the above argument, may have reached further into society.[68]

A core method of teaching literacy in Mesopotamia was by requiring students to copy out lists of words. About 85 per cent of the archaic Uruk texts we possess are administrative (mainly economic), but the balance (with almost 700 examples) are such lexical lists.[69] These lists may not only have served at the stage of instruction but also have remained useful thereafter as reference works for the literate, like our dictionaries.

The subjects of surviving lexical lists vary. A useful archaic list gives the names of towns/cities. Some list gods. Others list types of birds, or plants, or fish, or categories of wood and wooden objects, while others list produce and livestock of various kinds. One list (represented in many copies at Fara) had about 140 professions or occupations. In lists of officials, the titles represented widely different administrative roles – 'manager of xxx' – and such lists continued to be copied by students when they were already well out of date.[70]

Indeed, a curiosity of the lexical lists is that some of those developed in the Sumerian era persisted in educational use well into the later Babylonian kingdoms of the same region, and the Sumerian language was still studied after its practical use in society ceased. A comparison lies with the persistence of Latin at the core of school education in Europe into early modern times.

The Early Impacts and Roles of Writing 119

Egyptian scribal education used literary texts, as did that of Mesopotamia. But while there are earlier references to teachers and teaching, we only encounter the term for a scribal school later than the Old Kingdom.[71]

Literature

Modern Western urban society identifies 'literature' as a creative art form where the written form is privileged, to be read privately by the literate or read out loud to an audience. But the same Western tradition acknowledges that the greatest epics in the canon of European literature, the *Iliad* and *Odyssey* of Homer, were oral compositions which were performed and passed on in spoken form long before a written version existed. The same primacy of oral literature over the written form is found widely – from the Zoroastrian Avesta and the Qur'an to the Epic of Sundiata.

Where we find an early example in writing of 'literature' – something which does not fit another practical category – it is reasonable to consider it a tool of education for scribes and the elite. Using script to write and read the text of a poem or narrative well known and well established in oral form would be a valuable exercise; and literary works are still the core of modern teaching of ancient languages.

A text first seen in the Uruk III (Jemdet Nasr) period in archaic script continued to be copied in different edited versions through to the Babylonian period and represents the first example of what has been described as 'literature' – something quite different from the texts of lexical lists. Called the Tribute List (and alternately Sumerian Word List C), it interweaves combinations of figures and commodities with interspersed lines of repetition, giving the feel of a work of oral poetry.[72]

If we accept that the writing of literary work was part of an educational exercise, then the presence of this 'first work of literature' alongside the lexical lists removes the need to argue whether it is to be classified as an instructional work or a 'literary' text. An alternate view would see the evolution of written literary works as a formal institutionalisation of power by secular authorities, seeking to establish aspects of their rule and guidelines in the public sphere.[73]

Education of the elite involved more than just reading and writing. The term 'wisdom literature' is applied by historians to a broad range of texts, most typically comprising proverbs, common in the cultures of the ancient Near East, and is well known from examples in the Old Testament, the Jewish Tanakh.

The surviving copies of wisdom literature in the Sumerian language date from later than the era of Sumer and Akkad, by when Sumerian had the status of a dead classical language for study in the Babylonian schools of the second millennium, so the actual date of origin of such materials cannot be confirmed. A specialist scholar of such material included within the category of 'wisdom literature' instructional material such as moral guidelines (typically presented as those of a father to his son), practical instructions, other dialogues, school compositions, proverbs, riddles, short tales, fables, folk tales and satires.[74]

The 'Instructions of Shuruppak' is an early example of the genre. This text could serve both as a teaching tool in literacy and as an exercise to copy, while presenting practical rules for living which were equally appropriate in an educational setting. The first known example is from Abu Salabih, dated to the Early Dynastic III period in the mid 3rd millennium BC, ca. 2600–2500, and it became a classic work through Sumerian and Akkadian history.

Numerous guidelines given by a ruler Shuruppak to his son Ziusudra establish practical moral behaviours applicable in different life stages and different situations:

> You should not travel during the night: it can hide both good and evil.
> You should not buy an onager: it lasts (?) only until the end of the day.
> You should not have sex with your slave girl: she will chew you up (?).
> You should not curse strongly: it rebounds on you.[75]

The 'Kesh Temple Hymn' is another literary work whose origins trace back into Mesopotamia's Early Dynastic times, attested in tablets at Abu Salabih, but which continued to be copied and used in much later centuries. Such a hymn would probably have had its origins in temple rituals, with the words recorded for study and transmission. Fragments of other such hymns have been found among the early tablets.[76]

As the Sumerian city states developed, so did their literature.[77] Myths and epics expanded through the second half of the third millennium BC, with a poetic form indicating their probable origin in oral use. The famous Babylonian 'Epic of Gilgamesh' has its origins in tales originating towards the end of the Sumerian era.

Wisdom literature would also have been a feature of ancient Egypt. It is known to us only from Middle Kingdom examples, but some works may have earlier written or oral origins. The 'Instruction of Prince Hardjedef' might have its beginnings in the late Old Kingdom, as may the 'Instruction of Ptahhotep'.[78]

The Early Impacts and Roles of Writing

This is not to suggest that all literature in the ancient Near East was written for didactic purposes. Over time, more examples emerge. In Egypt's Old Kingdom, the Pyramid Texts have a literary style but a specific funeral purpose.

Social Class, Social Role and New Careers

We began this chapter by noting that writing emerged alongside the development of urban civilisation, craft specialisation and the state. Specialist craftspeople maintained their economic lives by the sale of their goods and services rather than their agricultural produce; merchants could buy and sell at a profit items manufactured by specialists or trade goods and produce at a distance from the area of production. This would produce disparities of wealth but not disparities of power.

The development of the state was marked by the growth of social classes. Those who ruled city states in Mesopotamia, and those whose local rule led to the unification of Egypt under a pharaoh, were an administrative elite with economic as well as political authority. The priests of Mesopotamian city cults were central to such power; the priests of pharaonic Egypt were part of the royal administration (with some related to the royal family too).

Writing could reinforce social distinctions. A division between the literate and the non-literate was largely one between those who could control economic organisation, political power and administrative complexity of the state and those who were subject to it. For those who were non-literate, a visual presentation complemented by a written form could achieve some of the same symbolic power.

Power, like wealth (and some technical skills too), passes through a family and is retained as far as possible by the family. But certain talents can facilitate upward mobility. The Chinese imperial examination system for public service could allow some to convert an ability to learn into an ability to rise in social status. European medieval and early modern institutions of learning assisted those more humbly born to demonstrate their talents and become figures in a religious, military, legal or civil hierarchy.

So we can construct an image of a scribal class of individuals – remarkably large in early Sumer – whose abilities and study brought them benefits within the structure of professions, classes and influence in the ancient world.[79] From the 26th century BC in Fara (Shuruppak), the name *Dubsar* signifies the profession of a scribe and promotes it in status lists. The education of scribes required able and no doubt specialist teachers who could bring their students in scribal schools to the required level effectively and confidently.

We can also demonstrate from the very beginning of writing on the clay tablets of southern Mesopotamia the emergence of that other literate figure, the bookkeeper (or accountant) who played such a crucial role in the relations between institutions and people in the exercise and recording of transactions concerning produce, land and more.

The invention of writing and its integration into the social, economic, administrative and religious life of early civilisations of the ancient Near East was the basis for other new roles and careers beyond those of the scribe, the accountant/bookkeeper and the administrator. The role of the surveyor was crucial in managing land allocations and transfers, developing public works and working with architects and engineers on major public monuments. The greater complexity of the surveyors' tasks meant that they needed to be able to record and transmit their measurements and their significance in definitive form: using text, measurements and visual presentations.

Finally, among new careers, was the changing role of the courier. Written messages between different sections of the civil administration, messages between military units and the beginnings of diplomatic correspondence all required reliable couriers to carry messages conveyed in a form that made their contents unambiguous: the written form. The role of the courier has been mentioned in Chapter 3, where horseback riders made a significant difference to distant communication. The significance returns in Chapter 6 when for the first time since the 3rd millennium BC, written texts ceased to be the necessary and such favoured means of communication as radio waves began to take over.

Propaganda and Pride

The power of the ruling elites of the ancient Near East was effected and maintained in a range of different ways. Functions such as canal and irrigation management emphasised the value of a coordinated central administration. Major public works funded from taxation could return revenue to subjects as payment for labour. Centralised markets enabled trade, and the larger the market centre, as developed in the early towns, the larger were the economic benefits. Security provided by the state was essential to make urban living safe; and the links of rulers and priestly class with the gods and goddesses gave additional security.

Knowledge of what the state elite could practically provide was therefore part of daily experience, but it was also strengthened by public relations:

The Early Impacts and Roles of Writing

propaganda. In this, and throughout the civilisations of the ancient world, propaganda benefited from a combination of public visual imagery and words.[80] The words might be a praise poem, or an oral testimony of a ruler's link to the divine. It could be a large relief inscribed with the royal titles and achievements, or an inscription on a visible object. Boasts of the power, achievements and services of the royal leaders were a priority in such propaganda, and those in the higher echelons of royal service would gain prestige by encouraging or organising such propaganda exercises.

Many of our ancient texts come from dedications to deities, where the monarch attests their loyalty or gratitude to a god or goddess. These may be intended to be visible in a temple or buried out of human sight. But even if they were buried, the very act of burying them would be a public scene and served to emphasise that the power of the ruler came from the support given by the divine. There was no divide between boasts to the gods and boasts to the population ruled. 'Display inscriptions' are found at Sumerian temples from 2800 BC and details of rulers from 2700. As with all propaganda, historical interpretation needs to distinguish between the claims of rulers' inscriptions and the reality on the ground.[81]

Eanatum, as king of Lagash, extended that city's power widely over Sumer. A victory monument shows military scenes on one side with a long inscription about victory on the other side.[82] Urukagina, ruler of Lagash in the 24th century BC, issued declarations, the 'Reform Document', outlining the changes he had brought about to improve the lot of his citizens. The document described how bad things were before his reforms:

> If sheep were bought, the influential man used to carry off the best of these sheep for himself ... If the son of a poor man laid out a fish pond, the influential man would take away its fish, and that man went unpunished.[83]

The final Sumerian ruler before the rise to power of Sargon of Akkad in 2350 BC, Lugalzagesi, king of Umma, commemorated himself with a lengthy inscription on a silver vase. In the form of thanks to the god Enlil, it boasted how his reign made life safe for the lands he had conquered, including Urukagina's Lagash.

Publicising the power of the king went well beyond the inscriptions placed in temple precincts. Military achievements were asserted in open-air sites outside cities. The Akkadian kings who ruled from ca. 2350 to 2200 BC, beginning with Sargon, used inscriptions to commemorate their deeds, including military victories, the building of temples and dedication of cults and the cutting of canals.[84] A major source for these, however, is from later Babylonian copies. Some were inscribed on statues or steles in the temple courtyard at Nippur to mark the start of a building project. Steles could also present their propaganda in purely visual form.[85] Akkadian citizens would use inscriptions on their seals to write not just their name, family and place but present images of some personal achievements.

Naram-Sin, the grandson of Sargon of Akkad, used writing in a definitively propagandist sense when he added the divine determinative *dingir* after his name in inscriptions; he not only respected the divine but promoted a comparable status.[86]

The later period of the Sumerian era, the Third Dynasty of Ur, is documented by texts which are known to us from later copies. Royal hymns from the period include both those in the voice of the king and hymns to the king, including identifying him as consort of the goddess Inanna: both serve the same propaganda role of asserting royal authority from divine approval.

The 'Curse of Agade' is a literary work, some 281 lines long, but it served to reinforce the values of the current monarchy by its denunciation of the supposed failure of a former Akkadian king, Naram-Sin, to honour Enlil and other gods. Although known only from later copies, it is thought to date from the Third Dynasty of Ur period:

> The life of Agade's sanctuary was brought to an end as if it had been only the life of a tiny carp in the deep waters, and all the cities were watching it. Like a mighty elephant, it bent its neck to the ground while they all raised their horns like mighty bulls. Like a dying dragon, it dragged its head on the earth and they jointly deprived it of honour as in a battle.[87]

Our main texts from Egypt prior to the Old Kingdom are mainly from funerary contexts rather than temple or residential sites. Finds from the elite and royal burial sites in Abydos and Sakkara carry the names of pharaohs, starting with Dynasty 0, represented in hieroglyphic form on carved objects and inscriptions. The names and title of the ruler are attached to demonstrations of power and authority.

Labels with complex combinations of imagery and lettering are found in the contents of Egyptian tombs. A perforated ebony tag (label) from Abydos mentions First Dynasty king Aha, and the goddess Neith, with scenes including a shrine and boats and an inscription in hieroglyphs whose meaning remains unclear (Figure 22).[88] With such complexity, the tag clearly served more than just to label an item, for which a simpler tag would suffice. Locations in tomb contexts suggests that such tags may exist to mark prestige by recording the position or achievements of the owner, perhaps used in life to promote their position. Then their use as grave goods extended this prestige into the afterlife.

The Early Impacts and Roles of Writing

FIGURE 22 Ebony label from Abydos with the name of King Aha (W. M. F. Petrie, Public domain, via Wikimedia Commons).

Senior Egyptian officials could use their tombs to proclaim their achievements on earth and their requests for a suitable afterlife. These resemble biographies, and we find actual biographic texts in tombs of the Old Kingdom, such as that of the Sakkara tomb of the official Hesy-Re in the early Third Dynasty (Figure 23).[89]

One of the best studied is an inscribed limestone block from the Abydos tomb of the senior official Weni (who became governor of Upper Egypt under the Sixth Dynasty), recording or claiming achievements in military command, public works and diplomacy, together with the honours placed on him by the pharaoh (including donations for his tomb).[90] Significantly, this information was placed not inside the tomb but on a tomb wall outside for all to read. Such a public declaration can be seen as a statement by a family of their status, which would reinforce their value to the pharaoh. It can also be seen as an assertion of the power and authority of the state – an argument reinforced by the deliberate damage to the location when the final Old Kingdom dynastic succession lost power.[91]

126   Developing Writing

When I was [master of the footstool] of the palace and sandal-bearer, the king of Upper and Lower Egypt, Mernere my lord, who lives forever, made me count, and governor of the South, southward to Elephantine, and northward to Aphroditopolis; for I was excellent to the heart of his majesty, for I was pleasant to the heart of his majesty, for his majesty loved me.[92]

FIGURE 23 Panel from the tomb of Hesy-Re (Djehouty, CC BY-SA 4.0, via Wikimedia Commons).

The Early Impacts and Roles of Writing

Unifying the State or Nation

A further function of writing deserves comment, a feature which, as argued in Chapters 5 and 6, was a major contribution of both the printing and the wireless revolutions.

Spoken language is the practical means by which people connect and which may define their identity. They do not need to use the same language as other groups with whom they share broad material culture, economic patterns of life, or to whom they are connected by trade networks. Non-literate societies, however complex, can coexist and interact without the necessity of speaking an identical language. Even with a language containing a broadly similar vocabulary, substantial diversity of dialects exists. Precolonial Aboriginal Australia, with trade and exchange networks across a vast landscape, had over 250 languages, and far more are known from New Guinea. The first printed books of European literature reached readers whose own languages and dialects varied widely and served to create a smaller group of national languages. The invention and spread of writing had a similar role.

Administration of a substantial state, like that of early pharaonic Egypt or a region of Sumer controlled by one or another powerful city state, needed an administrative system and administrative authorities, both civil and military, which could communicate through a common understanding. Civil and military officials widely dispersed needed to interpret not just the sounds of the symbols of writing but the words and meaning of those written symbols. So whereas predynastic Egypt could develop with linguistic difference between the Nubia of the Nile cataracts and the eastern Delta of the Sinai borders, a unified state needed a single language for communication. The education of scribes and officials in that language brought about unity in language, so that we can say writing did not just represent the spoken tongue; it unified it and created the coherent identity that language provides.

Religion and Words to the Gods

As Egyptologist Barry Kemp has emphasised, it is artificial to write of an ancient Egyptian (or Mesopotamian) 'religion' in the sense that the term is used of the modern Western world.[93] Others have drawn a continuity between religion, magic and medicine in the daily lives of the ancient world. The ancient civilisations had cults which required practical ritual; their populations also held world views on the relations between humans, divinities and the lived environment; between the living and the dead; and on their daily experience. We can tell specifically what people (and especially elites) *did* in relation to their tombs, their temples, their deities – practical aspects of that broad category we now call religion. We are less confident to say what the different classes, regions and groups of people *believed* as their world view.

It is equally important not to generalise too widely: we should not take evidence from later periods of Egyptian or Babylonian tradition and assume its earlier application or relevance.

Statements or actions of worship served not just to seek and acknowledge the support of a god or goddess but also provide public propaganda of the unity of earthly and supernatural rule. The Sumerian Kesh temple hymn is thus a work of literature and an artefact of a religious cult, but its creation (and subsequent copying) can also be seen as demonstrating publicly the relationship between city elite and the divine.

One late Egyptian text (ca. 710 BC) is a copy of a document which, from its style, may have its origins back in the late Old Kingdom. This 'Memphite Theology' is a public statement of the relationship between the god Ptah, here given a creator role, and Memphis as his powerful city:

> Thus Ptah was satisfied after he had made all things and all divine words, (59) He gave birth to the gods. He made the towns, He established the nomes, He placed the gods in their (60) shrines, He settled their offerings, He established their shrines, He made their bodies according to their wishes. Thus the gods entered into their bodies, of every wood, every stone, every clay, every thing that grows upon him in which they came to be. Thus were gathered to him all the gods and their kas, Content, united with the Lord of the Two Lands.[94]

Sumerian texts such as those from Nippur in the Third Dynasty of Ur included works described as liturgical.[95] Such Sumerian literary texts interwove mythical stories of men and deities, and these narratives continued to hold respect and be used in subsequent Babylonian and Assyrian times.

Spells and Magic

Incantations, spells and prayers to the gods are the means by which humans seek to influence the gods' behaviour to favour human needs. In Egyptian

The Early Impacts and Roles of Writing

belief, they also played an important part in seeking to secure an appropriate life after death: a life which, for the elite at least, was expected to match that on earth.

Some of the earliest known spells are from Sumerian Fara and in style reflect an older oral tradition.[96] Personal requests to the gods by elite private citizens are a feature from Early Dynastic Sumer and Akkad. Most commonly inscribed on vessels, but also on statues and other objects, individuals dedicated the objects to a deity in the hope of securing continuing good fortune for themselves or, sometimes, a third party. The search for good fortune in early Egypt featured visual symbols on amulets.[97]

As with other forms of early literature, we can assume a long tradition of oral spells, but by the later Old Kingdom we have written texts in the pyramids of the Fifth and Sixth Dynasty pharaohs – texts which are ancestral to the famous Book of the Dead in Egypt's New Kingdom from the mid 2nd millennium BC.

The Pyramid Texts are inscribed on the walls of chambers within the pyramid burials of Pharaoh Unas and other kings and queens. They take the form of hundreds of separate incantations, in a style confirming their origins in oral tradition. Their purpose is to ensure the pyramid occupiers' resurrection into the sky among the gods, as well as specific aspects of well-being in the afterlife.[98] The texts also include the ritual recitations for the temple associated with the pyramids. In later times, senior officials would copy some of the same texts onto their own coffins. Even if we conceded that these reflect a religious belief system, their presence in the tombs and the indication they were chanted by priests associated with royal burials is further evidence of text and propaganda emphasising the power and divine links of the monarch and family.

> Re Atum, this Unas comes to you, a spirit indestructible who lays claim to the place of the four pillars! Your son comes to you, this Unas comes to you, may you cross the sky united in the dark, may you rise in lightland, the place in which you shine! Seth, Nephthys go proclaim m Upper Egypt's gods and their spirits. This Unas comes, a spirit indestructible, if he wishes you to die, you will die.[99]

We know less about divination – the reading of physical signs to foretell advice on the future – in recovered texts. While we have references in literary texts to Sumerian seeking of omens, written records of omens in Mesopotamia are only known a millennium later.[100] Similarly, the textual evidence on divination in ancient Egypt comes from the later dynasties. This marks an interesting contrast with China, where the earliest examples of writing to survive are those in divination records.[101]

Law and Political Regulations

The social contract in a civilised state required protection of the citizen in exchange for loyalty to the rulers. A city wall may help protect against attackers from outside; the law protects against depredation from within the community. In the 24th century, the Sumerian king Urukagina boasted of the reforms he had instituted for the benefit of his citizens, and his conqueror Lugalzagesi boasted of his care for the safety of his subjects.

We could consider the instructions from father to son, like the Instructions of Shuruppak, as precedents for the state legal codes which specify punishments for misdemeanours.

Dating from the mid 3rd millennium onwards are formal legal texts, typically dated by year names. Those from the Third Dynasty of Ur at the end of the millennium are of greatest value. Of particular interest and importance is the 'law code' of Ur-Nammu (ca. 2112–2904 BC), with parts found in the Nippur archive and with additional tablets elsewhere. Among over fifty topics, it specifies the punishment for individual crimes (death for murder or robbery, prison and a fine for kidnapping) but also sets some principles of civil law (payment by a man to his wife if he divorces her). Once a law was in writing with copies made and distributed, it reduced the randomness and uncertainty associated with crime and punishment.

The decisions made by Sumerian courts were also recorded in the permanent form of cuneiform texts on fired clay tablets. Among some 300 court records from the Third Dynasty of Ur period, especially from the large city of Lagash, are notary documents of sale, loan, marriage and similar agreements and records of lawsuits about property and land.[102] They have not survived in excavations of Egyptian settlements, presumably because they were recorded on papyrus and linen.

Correspondence

When scholarly studies have focused on inscriptions, temple archives, administrative documents and texts in elite and royal funerary contexts, we tend to forget what we know was a most important form of writing in more modern history: that of correspondence. In Egypt, the pharaoh and his court were the centre of administration, but he ruled a complex administration across a vast area of north-east Africa and, often, into the Levant of western Asia. Civilian officials and military commanders needed contact with the centre and with each other. Such officials might serve at some distance from their own families: the couriers who carried official messages

The Early Impacts and Roles of Writing

might carry a letter from an official serving in Palestine to his son serving in the royal palace, or to advise on the management of his privately owned properties. As the class of those who could read and write grew, so too grew the norm of correspondence. While private letters are known from the late Old Kingdom, the largest archives are from later sources. Especially valuable in providing an insight into everyday life are almost 500 letters recovered from Deir el-Medina, a village of the workmen who served in building the New Kingdom royal tombs at Thebes.[103]

Clay tablets with personal correspondence also developed in late Sumerian times, though it was more common in the subsequent Babylonian era. The Babylonian mythical tale *Enmerkar and the Lord of Aratta* stated that the king of Uruk had invented writing in order to send a message to his adversary in Iran, avoiding the limitations of a messenger's memory or eloquence.[104]

This does remind us of the importance of correspondence in international relations; but a letter could only improve on the words carried by a courier if both parties had scribes who could read and translate in a common language. Diplomatic exchanges, and eventual treaties, required the reliability of written documents. The famous Amarna archive from the mid-14th century BC New Kingdom of Egypt provides close to 400 clay tablets of correspondence between the pharaoh's administration and rulers or Egyptian officials in western Asia, mostly in a cuneiform script. The Hittite rulers of Anatolia and regions of Upper Mesopotamia around the same time were active in diplomatic correspondence.[105]

Memory

Innovations not only derive from the human brain; they can change the human brain. Physiologists and psychologists have presented arguments that the development of the internet has changed the way we think and operate in more than just a practical and convenient way. The ready availability of an online or voice-operated search engine to answer a question and provide information means we do not need to remember what we used to store in our memory. The ready availability of computer programs to store data makes it possible to be highly professional with limited memory skills. Tracking back a few decades, the pocket calculator changed the need for much mental arithmetic.

The biological effects of artificial intelligence are more specific and have been under investigation and discussion for some time. A 2019 paper in the major international journal *World Psychiatry* summarised the research on the impacts of the internet on changing our cognition:

> Specifically, we explore how unique features of the online world may be influencing: a) attentional capacities, as the constantly evolving stream of online information encourages our divided attention across multiple media sources, at the expense of sustained concentration; b) memory processes, as this vast and ubiquitous source of online information begins to shift the way we retrieve, store, and even value knowledge; and c) social cognition, as the ability for online social settings to resemble and evoke real-world social processes creates a new interplay between the Internet and our social lives, including our self-concepts and self-esteem. Overall, the available evidence indicates that the Internet can produce both acute and sustained alterations in each of these areas of cognition, which may be reflected in changes in the brain.[106]

Practical aspects of reliance on the internet may affect memory use even in older citizens. The multiple distractions of email, messaging and social media may impact the working patterns of those established in their profession. But for those 'born digital', whose involvement in information is dominated by online sources, the long-term changes may be even more profound, and not just for those adolescents with internet or online gaming addictions. For some years, the leading British brain-research scientist Susan Greenfield has been promoting publicly (and somewhat controversially) her interpretation of the results of research on the changing brain.[107]

But even without the physiological changes to the operation of the brain, we can be certain of the many ways in which the internet and applications of artificial intelligence have altered our daily lives and the way we work, think, act and interact.

The arrival of writing, when before there had been only words and memory, would have had a similar impact in those sections of the community which became literate, and those who were impacted by literacy. Some of these have been described in this chapter. Laws could now be recorded and defined in writing, not just reliant on the opinions and decisions of judges and officials. Literature was learned and copied in schools, not only an oral performance with works committed to memory by performers and orators and transmitted by learning. Agreements in writing had a level of certainty that avoided dissent on who said what in verbal contracts. Memory and words were still important, but writing removed the need to remember in many social contexts.

Conclusion

This perspective was acknowledged by the 5th century BC Athenian philosopher Socrates. In Plato's *Phaedrus*, Socrates discusses the impact of writing with his pupil, citing (and perhaps creating) an Egyptian myth that the god Theuth (Thoth) had invented writing, alongside much in mathematics and astronomy; but it was writing that was suggested as controversial. Theuth claims that writing will make the Egyptians wiser and will improve their memory. But the Egyptian king has reservations:

> It will introduce forgetfulness into the soul of those who learn it: they will not practice using their memory because they will put their trust in writing, which is external and depends on signs that belong to others, instead of trying to remember from the inside, completely on their own. You have not discovered a potion for remembering, but for reminding; you provide your students with the appearance of wisdom, not with its reality. Your invention will enable them to hear many things without being properly taught, and they will imagine that they have come to know much while for the most part they will know nothing. And they will be difficult to get along with, since they will merely appear to be wise instead of really being so.[108]

Writing may have brought many benefits among its impacts, but as with internet search engines and more, its impact on the human memory may have been more nuanced and complex.

Conclusion

Was writing a gradual development, or an invention which subsequently evolved? Debates on this will continue.[109] It is relatively easy to consider how symbols for numerals developed and their uses spread with ease. It is also not difficult to see how a community, making records on papyrus or clay, would accept the convenient shorthand of symbols showing in an abbreviated stylised form a loaf of bread to mean 'bread' or a head with horns to mean 'cattle'. This is a convention of aides-memoires, not a writing system. But it is a major revolution when a symbol (recognisable as a real object or not) means the syllable *len* in any word in which syllable may exist or may represent a whole word whose identity from other words using the same symbol is indicated by a determinative suffix. The idea of writing requires not only the creation of a system of such symbols but also the teaching of their meaning to the whole pool of users, for their meaning is not intuitive. And such a pool of users is not finite but spreads across a city and soon beyond. It seems reasonable to suggest the idea of writing was an invention somewhere in the Middle East in the late 4th millennium BC – most probably in Sumer in southern Mesopotamia in the late Uruk period – and that its spread with identical meanings across the Mesopotamian region showed its efficacy and appeal; while it was the idea but not the characters that were taken up in western Iran (Elam) and in Egypt's Nile Valley. When Sumerian cuneiform was applied to the very different Semitic Akkadian language, its ability to transfer its syllabic meaning was invaluable. Alphabetical writing would eventually show the same transferability.

If the development of advanced agricultural societies in the Near East into urban civilisations can be said to represent a revolution, writing can be described as part of that revolution. The many applications and effects of the development of writing, as outlined in this chapter, provided tools but also a catalyst in transitions. Writing contributed functions to the changing needs of trade and economy and plays innovative roles with the emergence of new patterns of social class and power. In the third millennium civilisations of Sumerian Mesopotamia, and the Early Dynastic and Old Kingdom periods of Egypt, writing established its role in contexts which would continue in subsequent civilisations and subsequent millennia.

Literacy can provide power and authority: throughout history, the power of the pen has matched the power of the sword. Literacy can divide the world into two classes. Yet the written word has limited potential as propaganda: in a large visual display presenting powerful kings, influential gods or defeated enemies, it is the visual that has the widest impact, since the accompanying inscriptions can only be accessed by the literate minority. The power of propaganda – as with the medieval Christian church – may lie with those who can select, present and interpret the written word to those unable to access it themselves. Literacy creates a bond among those who can read and write: it enables civil, military and religious leaders to communicate in a code that is unintelligible to those under their control, command or influence. Learning to read as a marker of elite status was a stimulus to assist joining such an elite. But the wider spread of reading ability would await the stimulation of the printing press, as discussed in the next chapter.

One further transition is marked by the introduction of writing: that which enables us to study the past in new way. Our terminological distinction between prehistory and history may be porous, somewhat artificial and dated.[110] But there is a significant difference between the way that archaeologists and prehistorians study societies of which there is no written evidence, and those for which textual documents we can translate

Conclusion

complement all other sources of evidence. The availability of writing that has been deciphered and translated can bring different perspectives to our understanding. The impact of the invention of writing is therefore an impact on modern scholarship as much as on ancient society. But it also brings the risk of bias. If literacy was a feature only of the ruling elite, and presented often in public documents that give their perspective, then we risk giving less attention to the majority of ancient society who could neither read nor write. Their stories always existed even if they were not revealed in written words.

CHAPTER 5

Inventing Printing

> The wits and knowledges of men remain in books, exempted from the wrong of time, and capable of perpetual renovation. Neither are they fitly to be called images, because they generate still, and cast their seeds in the minds of others, provoking and causing infinite actions and opinions in succeeding ages.
>
> Francis Bacon, *The Advancement of Learning*, 1605

The Development of Printing and the Book

Although the digital age is well upon us, the powerful impact of print remains. Printed newspapers and magazines may talk of challenges to their survival but meanwhile the printed book thrives in ever increasing numbers. In the United States, a new book is issued by registered publishers every twenty-five seconds of the working day (in addition to over 2.3 million new self-published print and ebooks a year). Printed books still form a basic tool of education and the basic tool of religions and are among the most popular forms of entertainment while also complementing digital sources to provide information.

When we consider which innovations in human history had multiple impacts, the development of the printed book and other printed matter must come readily to mind. The ability to communicate an identical message across space and time, and to repeat that message as required, transformed aspects of the political, social and legal lives of those nations where printing first thrived. The Protestant Revolution, the so-called Age of Discovery across continents, the beginnings of new forms of scientific publication, the integration of written language – and much more – reflected the power of the printing press in early modern Europe, although despite the growth of literacy there were still marked divisions between those with personal access to the written word and others.

Woodblock printing of books had its origins in China, Korea and Japan, where it formed the basis of an active distribution of religious, secular and administrative publications, but long under close control. China was also the location for the invention of paper. But the development of a printing press with movable metal type that began in Germany was truly revolutionary.

In the medieval world of Europe, the written word was in manuscript: books accessible to the Church and to those with libraries (notably the institutions of religion, the institutions of the state, universities). These long had the power to interpret, censor and limit what was available to a wider audience. Europe's printing revolution in the middle of the 15th century allowed increasing numbers of literate people to access those words directly.

The innovation of printed books spread rapidly and stimulated the process to democratise knowledge as the late medieval world transformed into the early modern. Some 8 million copies of books in 30,000 editions were produced before 1500 by 1,000 printers in 300 towns across Europe. While different changing technologies in manufacture were affecting economic life, war and political balance, arguably the most important was the development of the printed book, because it led to changes in people's minds. Printing was a fundamental tool as a new generation of preachers established the beginnings of the Protestant movements in Europe. It aided the humanist movement in education and learning by making classical works more readily and widely available. It linked the science and technology of practitioners with those whose interest in science were primarily scholarly and theoretical, to the benefit of both. It gave a broad popular audience access to new and old literary works. Printing helped standardise national languages, as books in the vernacular came to supplement those in Latin, and authors and printers sought to reach as large an audience as possible. The educational textbook was developed for students. Printing meant that the same words in the same edition could be read simultaneously by many people, and to be read in different places, countries and settings. A book banned by one authority could be printed elsewhere and sold across borders. And in the new commercial enterprise of printing, the entrepreneurial printer could balance the risk of a new book with smaller print tasks such as indulgences for the Church, propaganda for the state or popular leaflets on a new discovery as Europe expanded its knowledge of the wider world. The format of the printed book, the technology of type, ink and paper and the new genres and audiences for books were established in just a few decades.

Despite our increasing knowledge of earlier printing in East Asia, the history of the printed book is conventionally tracked back to 'the Gutenberg revolution' in the mid-15th century, and the developments associated with his name of a printing press with cast and transferable metal type applied to paper and bound into volumes in print runs of books which could reach a wide audience across Europe. The burst of printing and publishing that followed served to emphasise the appeal and importance of the printed book. The early decades of print went alongside dramatic changes in European culture, religion and politics, and dramatic impacts in world history elsewhere.

This stimulates major questions. How much did the transition of books from manuscripts (available to the few) to print (available to many) influence the changes in society, and how much did it develop as a response to changes? And was the development of the technology associated with Gutenberg an absolute landmark, or was there a more complex history of the development of the printed book?

The traditional if simplified narrative is that printing in Europe began specifically and dramatically with the press established under the management of Johannes Gutenberg in the Mainz of the 1440s (a small German town of 6,000 people). This does give us a useful point in time and place against which to measure technological change and social movement. The awareness of a deeper history of printing has emerged vigorously over recent decades, not least in the description of earlier activities in China and Korea as important forerunners of the printing revolution, moving away from a Eurocentric view of history.

Printing in China

It is widely acknowledged that printing appeared first in imperial China, where a sequence of developments led to a large commercial market for printed books. But the technology was different from that seen in 15th century Europe, which focused on the flexibility as well as the advantageous economics of movable (and therefore reusable) type. The primary reason lay in the differences in writing. A typeface of fewer than 100 separate sorts could suffice for printing in a European language: upper- and lower-case letters, numerals, spacers and punctuation marks. Cast-metal founts for a typeset book could readily be distributed back to boxes for the next typesetting job. By contrast, Chinese books required tens of thousands of

The Development of Printing and the Book

individual characters, making it appealing to create a printing forme for single use, and in practice to do this as a woodblock, rather than expect to create, store and restore individual characters. Innovations in movable type never came to dominate Chinese printing.

The concept of printing in China can be traced back to the use of a seal to mark commercial products and documents. Other methods of reproduction included inked rubbings of stone inscriptions and the use of stencil to apply a pattern to fabric.[1] Woodblock printing developed with especial appeal in China, as elsewhere in East Asia, to impress repeat patterns on fabrics. It would be logical to extend this idea to apply woodblocks to paper which by the 3rd century AD was in common use in China for other purposes.

The Tang dynasty empress Wu Zetian, who reigned in China from 690 to 705, was eager to fulfil her religious loyalty to Buddhism by distributing large numbers of identical texts of a 'Great Spell' or *dharani*, produced by a mechanical process to bypass the labour and time of copyists. This pious act for talismanic purposes was commissioned towards the end of her life. A willing monk assisted in the project, which used woodblocks carved with the requisite lettering. But thereafter, imperial and religious prejudices seem responsible for slowing the development of woodblock printing of books.[2]

Other forms of printing would develop after more than a century. In 835, an official asked the Chinese emperor to ban the commercial printing of calendars because these were competing with the official calendar.[3] Such an item would have been a single page printed by cutting the lettering on woodblocks. A *Diamond Sutra* bearing a date equivalent to 868, consisting of a scroll made from seven paper sheets, was discovered in the Buddhist archive of Dunhuang on the Silk Road of western China.[4] This is the earliest known printed book to have survived but represented a growing industry.

Subjects for 9th-century woodblock-printed books included alchemy and astrology, medicine and dictionaries, as well as Buddhist religious texts. By the 10th century, more printing, both commercial and state-sponsored, was found in China. The Buddhist text of the *Tripitaka* found in Chengdu, printed in the late 10th century, is thought to have required 130,000 wood blocks. A 130-volume edition of Confucian classics was a major publication of the mid-10th century: a project with five woodblock pages prepared each day and twenty-two years taken to complete the exercise.[5] Sponsorship of publications was a matter of status for members of the Chinese elites.

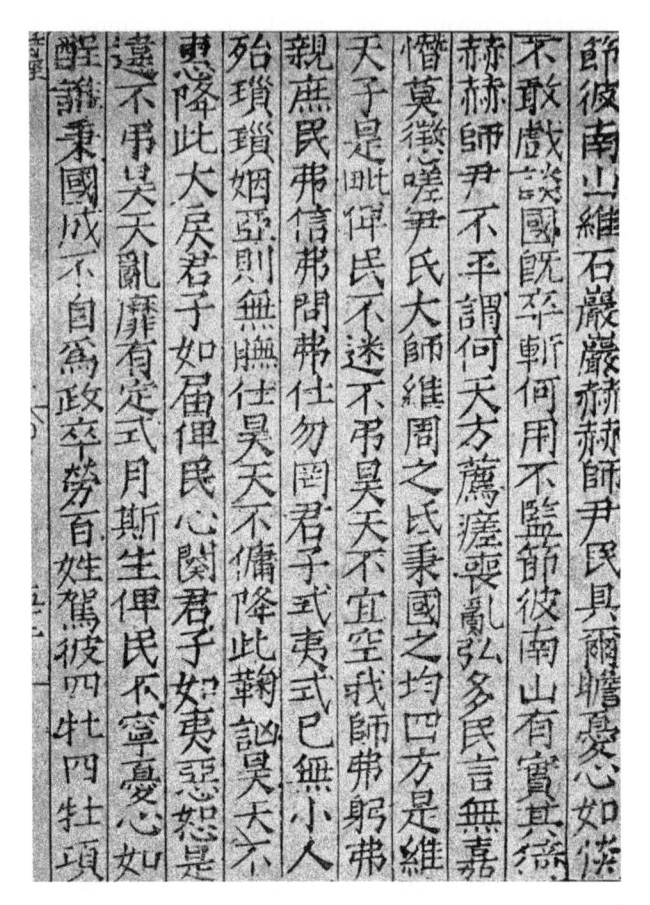

FIGURE 24 Chinese printing from the Song dynasty, 960–1269 (photo credit: Lebrecht Music & Arts/Alamy Stock Photo).

There was substantial expansion of printing under the Song dynasty, from the 10th to 13th centuries, covering a very wide range of genres, with the skill extending into other countries of East Asia. Printing and publishing continued under the Mongol rulers of China from the mid 13th to mid 14th centuries, and in that era major encyclopedias appeared. Given the vast area that fell under Mongol control, we can ask how far the *idea* (rather than the practice) of printing may have spread at this time.

Popular fiction was expanded in the Ming era after 1368, thus continuing to anticipate any European printing. The Yongle Encyclopedia completed from 1403 to 1408 – when Gutenberg was still a young child – comprised 23,000 chapters in 11,095 volumes with data drawn from 7,000 earlier

The Development of Printing and the Book

publications.[6] Another developing category was educational textbooks. A bibliography of print publications from imperial China exceeds 250,000 titles.[7]

It is notable that the use of printing in China did not in itself appear to revolutionise society or have dramatic transformative effects on politics, religion or culture, as would be seen in Europe. Printing fulfilled religious, administrative and educational roles endorsed by China's rulers; it did not provide challenges to the status quo. But printing did spread knowledge and produce a large educated elite.

A typical pattern of Chinese woodblock printing was to write out the required text on a waxed sheet of fine paper, which was then impressed on a wooden block bearing a thin layer of rice paste. This negative transfer of the characters required for printing meant an engraver could then cut out the block to leave the required letters in relief.[8] Such wooden blocks could print many copies, whether for a single sheet or the pages for a lengthy book.

Movable type – the feature which made Europe's printing revolution so cost-effective – did have early examples in China, but it did not replace the woodblock.[9] Experiments with type made from clay and then fired began in the 11th century with limited take-up, although examples of clay type are known from a few regions in the 13th century.[10]

The first trials using wood type also dated from the same time as the first in fired clay. Movable type in wood was in use from the early 14th century, an initiative of Wang Zhen (Chen) of Shandong on China's east coast. This required 30,000 different pieces, with characters held for selection on a revolving table and organised by rhyme.

A minority of books were printed thereafter using the movable wooden-type method, with its use favoured especially by the imperial court and some local ruling families for their own book projects. The cost of creating a set of wooden type was high, and so once made it maintained its value. In Chinese culture, it could be said that wood was considered the natural printing medium: it was therefore a choice between woodblock and wooden type, without an incentive to adopt other type materials.[11]

Wooden type found on the Silk Road at Dunhuang served to set letters in the Uighur script, and the type includes combined characters (ligatures) as well as simple characters.[12]

Meanwhile, some movable metal type was developed in select areas of China from the later 15th century, by when it was in wide use in Europe; but it was bronze, rather than lead or a lead alloy, which was favoured.[13] Tin had been tried but was found to give poor results with the kind of ink used in

China. It is not thought that European movable metal type influenced this Chinese development, which followed only shortly later; and Chinese metal type may have been laboriously hand cut rather than the economic mould casting of Europe. One major publication printed in metal type was an 18th-century encyclopedia, but with the fount melted down afterwards.[14] Woodblock printing survived and thrived after metal type printing had begun in Europe, with the first metal type for Chinese characters coming from European-owned sources in the 19th century.

Given that a complete set of type to cover all Chinese language might require 200,000 characters, the survival of woodblock printing into the modern era is not surprising. Manuscript copying remained active and of high status in imperial China. The cost difference between creating a manuscript copy and carving a woodblock to print multiple copies was certainly real but was less than the difference the Gutenberg revolution made to European books, once the costs of binding were included. While Chinese book printing was a substantial industry before and after the era of Gutenberg, it was not one transformed by movable type until the later 19th century.

Printing in Korea and Japan

The pattern of using woodblocks in printing was known very early in Korea and also in Japan, although later than in China. Woodblock printing of fabric had been well established before its application for text.

A scroll from the stone pagoda at Bulguksa Kyongju in Korea, carrying a Buddhist sutra in Chinese, was printed from twelve woodblocks at some point before the year 751. However, this may have been printed in China rather than within Korea. By the 12th century, local Korean printing was well under way.[15]

A Korean text from 1234 refers to twenty-eight copies of a book printed in 'metal letters'. A book printed in Korea in 1377 may be the first known from metal-cast type.[16] Bronze-cast letters are better attested from the 15th century, and movable type maintained its importance in contrast to China; an official 'Bureau of Type Casting' was established in 1403, with copper cast in clay moulds, although these clays would not survive for a second casting.[17] Different metal alloys were used over time. A vigorous market for printed books developed, although the books approved for printing were limited largely to Confucian texts and official government publications.[18]

Despite this, the printing press did not develop to take advantage of the cast founts, with a much slower process of laying paper over the inked metal

The Development of Printing and the Book

type and applying pressure to transfer the image.[19] Korea has a somewhat closer parallel than does China to the German development of metal movable-type books, but Korea is even further from Europe than China and the Silk Road.

Paper books were known in Japan before they began their own paper manufacture, probably by the 7th century. The empress Shotoku ordered the production of a *dharani* in the 760s 'in a million copies'.[20] While the number would be symbolic, the idea of multiples of the same Buddhist text reproduced by mechanical means (and therefore presumably woodblocks) is remarkable. The *dharani* were intended for deposition rather than specifically to be read. Despite (or in some interpretations, because of) the empress's initiative, printing faded from view in Japan for another two centuries.

Complete printed books are known only later in Japan. One thousand copies were printed of the Lotus Sutra in 1009.[21] Expansion under Buddhist religious guidance was seen from the 11th century onwards, but using Chinese characters; Japanese script appeared in print from the early 14th century. Some Chinese woodblock carvers moved to work in Japan. Movable-type printing only dates from the secularisation of Japanese printing in the early 17th century.

Invention of Paper

However we may judge and doubt the possibility that the 'idea of printing' may have reached Europe from Asia, it was paper that made Europe's industry of printing with movable type possible; and the invention of paper was a Chinese achievement.

While the ancient civilisations of western Asia had used cuneiform impressions on clay to record their texts, ancient Egypt developed papyrus as a writing material and developed a sophisticated chemistry of inks.[22] The plentiful papyrus plant lent itself to the production of inexpensive writing materials, which played their part in communication from the 3rd millennium BC and through the Classical world of Greece, Rome and Byzantium to the beginnings of medieval Europe. Longer documents were written either on scrolls or later as pages cut and bound into a codex. While the use of papyrus was well suited to arid regions, it lacked longevity in the damper climates of Europe.

Well established by the later Classical era, writing on parchment (treated animal skins) or vellum (the more refined version made just from young animals, especially calfskin) had a durability that was better suited to variable climates, and this durability gave it value and importance in documents intended to last. These might be official statements and certificates, or they might be books for the religious, university and private libraries of Europe and the Islamic world. While some book collectors would continue for some time to value new manuscripts or even printed books on parchment or vellum after the introduction of paper, the inauguration of book printing would eventually bring about its demise.

The writing traditions of China had used different materials. Tablets of bamboo and wood were the traditional materials on which Chinese texts were written, continuing in use until the 3rd or even 4th century AD. Meanwhile, silk, another Chinese discovery, began to find favour for writing, valued for its lightness and durability. Writing on silk is known as early as the 7th or 6th century BC and continued to the 3rd or 4th century AD, overlapping with the use of paper.[23]

The Chinese invention of paper, made from disintegrated and processed vegetable fibres, or cloth rags, probably dates as early as the 1st century AD. It was not developed initially to carry writing, but for wrapping and other decorative purposes. Of course, the multiple uses of paper in Chinese culture have remained into modern times: not least the representation of items in offerings, and the long-established use of paper as money. By the 3rd century AD, paper was widely used in China for writing, then printing, and that use spread to elsewhere in East Asia. The fineness of Chinese paper put limitations on its suitability for printing on a press.

Traders from the Islamic Middle East seem to have been responsible for conveying the idea of paper and the methods of paper manufacture to their own cultures, after Arab traders encountered it in Central Asia.[24] Paper was being manufactured and used for writing in south-west Asia by the 8th century, where it had a quick take-up. Chinese craftsmen worked in 8th-century Baghdad to create paper for the local market and make the city a centre for supply, and Arabs thereafter developed their own methods of paper manufacturing, favouring textile rags over plant fibres. One appeal of paper in official usage was that once written on, the ink could not readily be removed: it was the most reliable means to avoid forgery.

It took longer to challenge parchment in Europe, but it has become clear that paper had a significant presence in medieval Europe well before the application of printing.[25] European travellers to the east had seen the uses of paper for money but initially paid less attention to it as a writing material. The knowledge of paper manufacture as modified and developed in the Arab world spread from the Muslim world of North Africa into Spain by the 11th century, and paper documents were known in Italy by the early 12th

The Development of Printing and the Book

century. A prejudice in favour of parchment remained: a decree from the authorities in Sicily ruled against the use of paper in 1146, and a similar ruling by Frederick II of the Holy Roman Empire dates from 1231, parchment being seen as the appropriate material for official documents.[26] The Italian paper industry was established by the 13th century and that of Germany and France by the 14th. Chaucer used 'paper white' as a means to define a pure colour in his *Legend of Dido*, implying its familiarity to his readers. Arguably paper's impact awaited the 15th century: it was the development of printing that required paper for mass production of books, and it was printing that stimulated the production of paper; now with printing the paper mills had a dramatic and expanding new market.

Scribal Copies

The creation of books was a manual skill – and a specialist skill – before the emergence of printing in Europe. The Islamic world had a dedication to copying works of religion, history, philosophy and science, and this extended to translations of classic Greek works. Such a tradition in manuscript Arabic and Persian texts would remain as the Muslim world long resisted the adoption of printing technology.

Christian manuscripts were at first specifically the product of those in monasteries and other religious institutions, where the copying of text and the illustration and colouring of manuscripts was a core activity and an act of piety. One source from a 9th-century European copyist of manuscripts noted it took thirty-four days to create a manuscript book at the rate of eleven pages a day.[27] The scriptorium within a monastery might have one person dictating while a second inscribed the text, a system which may have helped avoid omitting lines but permitted inconsistent spellings.

Different scribes might work simultaneously on separate parts of a text to speed its overall completion. And as the demand for texts continued to grow in the later medieval period, those in religious orders (nuns and church clerics as well as monks) could be complemented by lay scribes. Northern Italy developed as a major centre of scribes and illustrators to fill requirements of the market, even though the demand was dominated by religious works.

Commerce did operate alongside piety. The monastery of Tours developed a line of bibles in a standardised design for sale through the Carolingian Empire of the 9th century. Mass books were produced and sold widely by the abbey of Saint Amand.

Growing demand for books in the later medieval period, including the development of universities in Europe after the late 11th century, stimulated the emergence of commercial copyists outside of the scriptoria of the religious institutions. This had other effects: as long as the institutions of the Church had been the producers of manuscript copies, the Church authorities could more directly control what was copied.[28] Now the development of commercial copyists reduced this direct control, and market forces came into play.

Nevertheless, the Church continued to be a large customer, and religious books still dominated numerically: bibles, missals and books of the liturgy, reference works of canon law, works by the early Christian Fathers and biblical commentaries.

In addition, editions of classical Greek and Latin authors were needed for teaching. Outside of monastic institutions, lay copyists created books to be sold, or hired out in parts for students to copy for themselves, through booksellers (or stationers) who were notably established in university towns. Copyist fees were recorded: different fees for a book copied with fine script for an institution or library and books copied more quickly into a cursive script for practical use.[29] Additional fees would be payable to the illustrator/rubricator and to the bookbinder.

The growth of education and literacy in the early 15th century increased the market for a textbook, a Book of Hours and the like, which stimulated manuscript production further. The reading public continued to expand as romances and popular writings complemented books of a religious nature and classical works, and books of medicine and law were essential to the workings of society.

The potential popular market that would be met by printed books was presaged by mass copying such as that in Alsace in the first half of the 15th century, where a production line of copyists and illustrators provided a range of texts from romantic epics and prayer books to legal texts.[30] The substantial growth of manuscript production and the shift to workshops of lay copyists in the 14th and earlier 15th centuries provided an incentive and reward for the development of printing: a drive to find a new technology to meet the growing markets. A full-length book might take a scribe a month or longer to create a single copy; once typeset, a printer could create a full print run in a week.

There was a backlash from defenders of the manuscript. In 1515, Abbot Johannes Trithemius protested that books printed on paper would never last as long as manuscript on vellum or be undertaken with as much care.[31] This argument has some resemblance to anxious concerns on the future

The Development of Printing and the Book 147

inaccessibility of our contemporary electronic-only publications. Manuscript sales continued beyond the 16th century, as some of the wealthy saw value in unique copies for their libraries.

Precedents for Printing in Europe

'Printing' in the sense of making marks on a surface that can be repeated was nothing new for Europe. The commercial seal as used through the ancient world of the Middle East (including the ubiquitous cylinder seals of clay) had continued in the Classical civilisations of the Mediterranean. Dies and stamps had long been used for coins, since their value lay in their carrying a recognisable repeated image (typically the head of a ruler) and inscribed lettering on a metal disc of standard size. Printing of repeated patterns on cloth arrived much later.

There are parallels between Chinese woodblock methods and those adopted by Europeans for purposes other than the text of books before Gutenberg. European experiments with woodblock printing to produce multiple copies did not long precede movable type, starting perhaps seventy years before the introduction of the printing press. By the earlier 15th century, woodblocks were in use to print playing cards and also to create portable religious images for the faithful. Once metal type could be set in the new technology, woodcuts (carved blocks less than a full page) were still required to add in the illustrations. Woodblocks were also used in early printed books for initial letters of chapters to emulate the lavish presentation of manuscripts.

Since early printed books after illustration and binding were still expensive, woodblocks of text were developed over the second half of the 15th century in an attempt to compete. Manual labour in cutting the blocks of type had the modest advantage that illustrations were carved into the same printing plate as the text. It has been suggested that since print runs for most early books printed by movable type were modest, woodblock printing could technically have managed to produce similar print runs.[32] Block books overlapped with movable type for some sixty years, and carved images in wood for reproduction would continue much later than that.[33]

The European Printing Revolution

As with paper, Persian and Arab travellers encountered book printing ahead of any European development, yet the Islamic world did not adopt the practice.[34] The Mongol Empire, which from 1260 to 1368 controlled China (where they sponsored printed books), spread far to the west, and Chinese printing was described by Persian and Arab travellers. Within the Islamic Middle East, hostility to printing until the 18th century reflected in part religious attitudes to the physical text, in part admiration for written calligraphy, despite the substantial scholarly traditions of Islam.[35] A small amount of printing is known from Egypt dating between the 10th and 14th centuries: fragments of religious texts on single sheets of paper.[36] In the Ottoman Empire, non-Muslim communities printed their own works: a Hebrew volume from Istanbul in 1503–1504, and some Christian publications from the 16th century.

Travellers from Europe to China and other areas of Mongol possessions in the 13th and 14th centuries would have seen evidence of book printing, but they were particularly impressed by the concept and use of printed paper as money, and their writings reported this with wonder.

Already in his 1586 book on the history of China, the Spaniard Juan González de Mendoza suggested that the European introduction of printing owed its origins to China with the knowledge passed by traders between China and Europe via Russia or by sea.[37] Many authors have since emphasised the Chinese innovation ahead of that in Europe; some consider China to have inspired or otherwise influenced Gutenberg, but no evidence exists to confirm any flow of technical information between China (or Korea) and the German innovators. Because of the differences in method, it is possible (though far from certain) that 'the idea of printing' rather than any technical knowledge passed along the Silk Road, and the innovative methods of European printing owed nothing to the east.

European book printing with movable cast-metal type began in Germany, in Mainz, or perhaps a little earlier in Strasbourg. That weakens any view that the idea of printed books was due to a westward spread from China via its neighbours. Europe's first metal-type printing was not in Constantinople, not in Venice, not in Genoa, where links with trade to the East might have inspired local entrepreneurs. One of Gutenberg's biographers advanced the hypothesis that the Catholic priest (and later Cardinal) Nicolas of Cues, who may possibly have known Gutenberg, did visit Constantinople in 1437 and could have heard there of Korean printing.[38]

Movable type was possible, attractive and affordable because of a script that was written in just a few letters: twenty-six letters would cover most western European languages, though printing required upper and lower case, ligatures to represent combined characters in roman script, numerals and punctuation. By contrast, Chinese writing with many thousands of

The Development of Printing and the Book

characters could use woodblocks to print but would lack economies of scale to reuse these.

The major features we associate with European printing were developed together, and the new technologies made such an economic impact that the technology spread rapidly in Europe. Indeed, the methods and details of the Gutenberg innovations were such that they remained with limited change for 350 years. Book printing developed and integrated what had previously been different skills. Metalworkers had made a punch for impressions; casting was required for coins; wooden blocks had been inked and used to impress images and often accompanying words on single sheets.

For the new industry of printing with movable type, a hand mould for each letter was required first – made often of copper and based initially on the best calligraphy of contemporary manuscripts. Then, a molten metal (typically an alloy of lead with a small amount of tin, antinomy and possibly other material) was poured into the mould, which could be used repeatedly to cast type. The cast type would require hand finishing: while the height of the letter was set by the accuracy of the mould, the foot of the type would need planning to a standard height for use in the printing forme. One calculation noted that since 18th-century typefounders were able to create four types a minute, even 15th-century craftsmen could be highly productive in the use of the pioneering moulds.[39] The cost of each new piece of type was therefore modest.

The cast types were stored in readily available compartments, ready to select and insert in a composing stick. A galley of variable height stood ready to receive the typesetting to make up pages with strips of blank 'leading' to space out the lines of text. By distributing the type back into composer's boxes after each page was printed, there were no further costs in metal or labour for type. And once a type had been acquired, a significant cost of a book lay in paper, so establishing a print run that was saleable was important to avoid capital being tied up in stock.

While the cost of paper meant too large a print run for the market was a commercial problem, so too was a failure to print enough copies, since type was returned to the typesetter's box for reuse as soon as a book's pages were inked. New demand outstripping existing stock meant the whole typesetting exercise had to begin from scratch.

The further economy lay in the printing press itself (Figure 25), a development of what was known from block printing of fabrics. The printer's skill lay in ensuring the correct condition of the paper, adjusting the even height of type, using the appropriate level of inking and applying

FIGURE 25 Sixteenth-century German printing press (Conradus Sweynheym & Arnoldus Pannartz, Public domain, via Wikimedia Commons).

the right pressure of the platen to bring the printed impression onto the paper. The sheets of paper (or vellum) had to be exactly positioned for printing accurately on two sides. In time, multiple pages would be printed together before the sheets were folded into the printed sections for binding.

Coloured letters and headings – to match the appearance of those in manuscripts – were added by hand. Illustrations used woodcuts in the first stage of book printing. Those whose employment had been in illustrating manuscripts might find new employment in the first decades of printed books. Such labour would add to the already expensive costs of a major printed work such as the early printed bibles but would reduce as printing served a wider secular audience in larger and therefore less expensive print runs.

Customers would expect their books to be kept in an acceptable and readily handled form – binding now, like in early printing, being a complex and relatively costly part of book production. Initially, customers for Gutenberg's Bible seem to have arranged their own binding.

The Development of Printing and the Book

Gutenberg and Mainz

We associate the introduction of movable type and the printing press with the name of Johann Gensfleisch zum Gutenberg from the German-speaking town of Mainz on the Rhine, better known to us as Johann Gutenberg (1400 or 1403 to 1468). His background had given him skills in metal working, and while in Strasbourg for ten years he extended his familiarity with the delicate metal-cutting and metal-working techniques which would be adapted to type. He seems to have developed there the ideas and skills which led him to establish a type foundry, typesetting and book-printing operation back in his home town of Mainz about 1448. The team associated with Gutenberg's printing press and printing works (including those who contributed financially to its establishment) are recognised as pioneers in the new industry. Although Gutenberg himself remains regarded widely as the originator of Europe's model of printing with movable type, others have shared this credit equally in some narratives.[40]

The first books to be printed were not biblical texts or religious tracts, nor administrative documents or books for general reading. They were educational textbooks: the *Donatus* (the *Ars Minor* by 4th-century Roman Aelius Donatus), which taught Latin to Europeans scholars in a Latin script, was the foundation of the industry.[41] But the first of these twenty-eight-page texts were printed on vellum for longevity (to pass through the hands of many users, reluctant or enthusiastic) – whereas paper was the norm for most future titles. The poem of the *Sibyllenbuch*, which in full would have been perhaps twenty-eight pages long, also seems to have been one of the earliest printings, in this case in German. This first fount included different widths of the same letter in order to assist towards creating lines of approximately even length, so 202 different characters (Figure 26).

A common historical narrative attests that the development of printed books served to erode the authority of the Roman Catholic Church hierarchy and usher in the Protestant movement. But despite that image, the Church was largely in favour of the printing revolution, which provided them with trustworthy religious books as well as indulgences to sell. The financial security of the first printeries in Mainz was underwritten by Gutenberg's production in 1454 and 1455 of large quantities (many thousands) of indulgences (typically on vellum rather than paper) for the Church to sell as fundraisers, in this case to support the campaign against the Ottoman Turks. These were the very letters of indulgence offering remission of sins that would inspire the outrage of the reformers whose

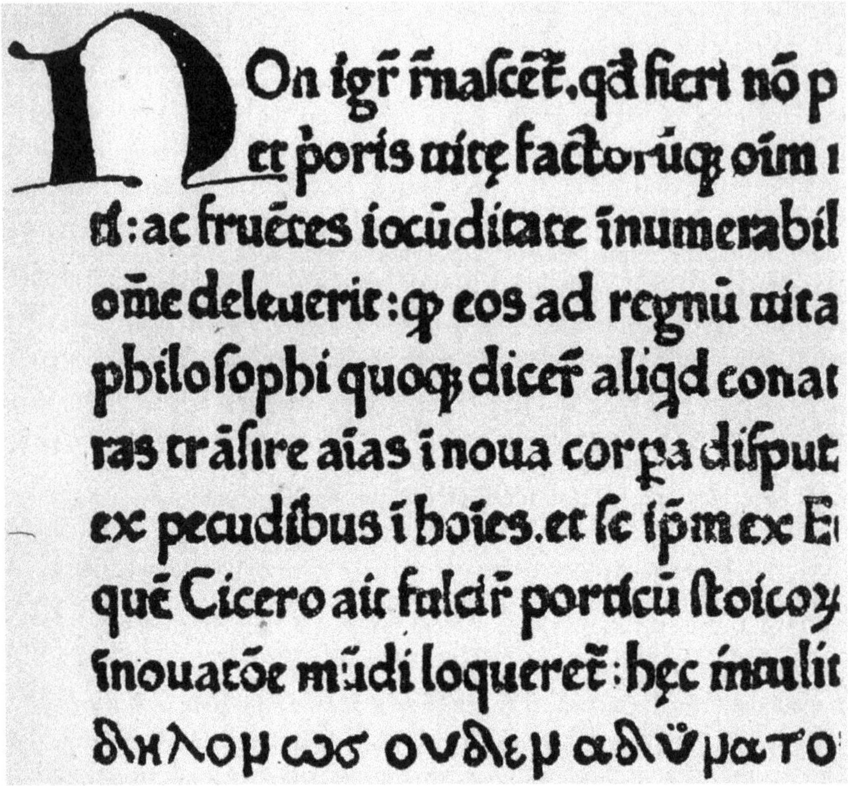
N On igr̄ n̄nascēt. qđ fieri nō p
et poris mirę factorūq; oīm i
ti: ac fruētes iocūditate inumerabil
oīne deleuerit: q eos ad regnū mita
philosophi quoq; dicer̄ aliqd conat
ras trāsire aīas i noua corpa disput
ex pecudibus i hoīes. et se ipm ex E
quē Cicero ait fulcir̄ porticū stoicoꝝ
inouatōe mūdi loqueret: hęc inaulit
δηλον ως ουδεν αδυνατο

FIGURE 26 Typeface of the Subiaco Lactantius (1465) (photo credit: Art Collection 2/Alamy Stock Photo).

denunciations and critiques led to the Protestant Reformation.[42] They were effectively blank forms on which an official church signature and date assigned remission to the contributing individual. The first fount developed for books was also used for the smaller tasks we call jobbing printing, such as calendars.

For printing the Bible in its then current Latin translation, Gutenberg's workshop developed a new fount of 290 types, to allow for upper and lower case, ligatures, punctuation, numerals, special sorts and spacing. This major publication, initially in an edition of 180 copies of 1,282 pages, set patterns that would long follow: a page of 42 lines with 2 columns on pages of 420 mm by 320 mm (16.5 by 12.5 in); many modern books average 39, on a smaller page. Right justification was achieved by spacing letters; with selected word breaks to maintain the straight right margin, and leading

The Development of Printing and the Book 153

between lines of text to ensure readability. There were copies printed on vellum (using the skins of perhaps 5,000 calves) and copies printed on paper (as a raw material this cost a third of the vellum). Hand colouring was added.

Completed in 1455, this Bible marked a new authority and visibility of printing technology (Figure 27). An estimate suggests 46,000 individual types would have been needed, with four to six compositors working on the project simultaneously. Gutenberg was probably responsible for a thirty-six-line Bible (also with two columns) dating from 1458 to 1460, which gave a longer extent for the finished work.

Gutenberg had opened two print shops, but a clash with his partner and investor Johann Fust led to a rival print shop in Mainz under Peter Schöffer. There was clearly more than enough immediate work for two printeries.

Religious books to serve the needs of the Church provided a flow of income, although differing markets might require different founts to cover Latin, German and in due course elsewhere other languages, in different type sizes to suit different contexts of use. The production of the Mainz Psalter including text of psalms and other material for use in church services marked a new quality of what could be produced by the new technology, with a printing in 1457 and new versions thereafter.[43] The

FIGURE 27 The Gutenberg Bible (Kpalion, CC BY-SA 4.0, via Wikimedia Commons).

Catholicon of 1460 was an encyclopedic dictionary and grammar of Latin produced in Mainz, some 744 pages in length with a new typeface again required.

But small print jobs such as the calendars continued, alongside other modest items. A new use of printing was the political pamphlet, with a mass leaflet produced in political conflicts of 1462.[44] Advancing a viewpoint by word of mouth and by handwritten posters may not have ended, but a new technique was now available for propaganda, subject to the agreement of the printer to undertake such work.

Political propaganda and political conflict meant military conflict and therefore winners and losers, and Gutenberg seems to have been identified with a losing side, leaving Mainz in 1462 to set up a new printery in the small town of Eltville. But this was only one in what was now a fast and substantial spread of the new industry.

The Spread of Printing

Given the economies of scale in production of books and other documents provided by the technology of movable type, it is not surprising that it spread rapidly in Europe, with new workshops established both by those who had learned directly from Gutenberg before his death in 1468 and by those who copied the techniques. Printing expanded from Mainz and Strasbourg to Cologne and Nuremberg and to Basle in Switzerland. In all, more than 300 print shops in 60 German-speaking towns opened during the first half-century of the innovation.[45] It was initially German printers who took the technology to other countries: Italy, France and beyond. The speed of its spread and growth is, however, remarkable. It is estimated that by the year 1500 there were more than 1,000 printeries in about 300 towns across Europe, and they had produced at least 30,000 editions of 10,000–15,000 different publications in total, with estimates of 8 or 9 million volumes or more, although print runs as small as a few hundred were common.[46] Printeries in just twelve towns were responsible for over two-thirds of what are called incunabula (in German *Wiegendrucke* or cradle prints), signifying books printed before 1500, a very fast growth from the cradle! By the next century, the productivity and expertise of the printing industry meant than a single press might print up to 1,500 sheets a day – with printing on both sides.[47]

In the 16th century, there were perhaps 150–200 million copies printed in over 200,000 editions: an estimated 79,000 editions before 1500 and 138,000

The Development of Printing and the Book

in the next half-century.[48] The Universal Short Title Catalogue has listed over 750,000 editions published in the first two centuries of printing.[49]

Venice was a major early centre, and it was there that the 'roman typeface' developed, a very readable fount which contrasted with the lavish lettering based on that of manuscripts.[50] In the 15th century, 150 Venetian print shops produced 4,500 titles. Paris was also an active centre of printing, followed by Lyon. Printeries were set up before 1500 in Spain, Portugal, Belgium, the Netherlands, England, Bohemia, Moravia, Hungary, Yugoslavia, Denmark and Sweden. Individual printers would move from city to city, helping entrepreneurs establish new presses. Founts would be cast from the same moulds and sold elsewhere, speeding up the rate of new print shops. The first print shop in the New World opened in Mexico in 1540. Of perhaps 345,000 publications before 1600, almost 84 per cent were printed in Germany, Italy, France or the Low Countries.[51]

Despite the growth of universities in the 15th century, university towns were not the first priorities for printers to set up their workshops. Other towns were more appealing, partly because they had a strong commercial basis with networks of trade so could act as a base for travelling salesmen and agents to take their wares off to university and other towns over a wide area.[52]

It would take some time for universities to employ their own printer/publishers. The publisher of this book, Cambridge University Press, can trace its continuous publication programme back to 1584, even though official permission to appoint printers had been granted fifty years earlier.[53]

Englishman William Caxton operated a printery first in Bruges, but his edition of Chaucer's *The Canterbury Tales* of 1476 came from an English printery he set up in London. The first book printed in Oxford dates from 1478. English printing was English-language printing (including translations from Continental works), intended just for the domestic market and reflecting commercial rather than religious roots, although commercial priorities meant that Caxton too printed indulgences.[54] Of an estimated 15,000 publications in England before 1600, 89 per cent were in English rather than the classical and international languages.

The indulgence market was huge and well serviced by printers, who could balance such smaller jobs for customers such as the Church that paid cash against the commercial risk of a printed book. Records include an order for 2,000 indulgences required in 3 weeks; a total of 143,000 indulgence supplied to a Benedictine monastery over a 15-month period in 1499–1500; and 20,000 'certificates of confession' from an Augsburg printer in 1480.[55]

The range also developed rapidly. The Church, of course, remained a major client, requiring alongside indulgences, bibles (including a German-language version in 1466), psalters, as well as missals needed by each local parish priest – no longer reliant on the maintenance of manuscript copies. While the missal was a major item in early printing, an even larger number of editions were printed of a breviary (*breviarium*), with the texts required for daily utterance by the Church. Also widely printed was the *psalterium* or book of psalms. Choirbooks were seen in far fewer editions.[56] Devotional books for individuals such as a Book of Hours intended for a lay audience might be commissioned by the Church for use by lay readers, and printers took to issuing their own versions in search of a commercial success, which was not always within their grasp.[57]

Educational textbooks remained, with Latin vocabularies profitable. Individual authors' books began publication for general sale, some augmented with woodcut illustrations – whether works of religious piety or of

FIGURE 28 Publication about Amerigo Vespucci's travels to the New World, 1504–1505 (Public domain, via Wikimedia Commons).

history, philosophy or legal subjects. A large market was discovered for works of classical authors which would suit student use, with Italian printers especially successful with this genre. Publication of more recent literature presented the broader range of languages. In 1484–1485, illustrated guides to herbal medicines, one in Latin and one in German, marked the beginning of what would later be considered as self-help works, so popular that they were soon copied in unauthorised editions (for of course legal copyright did not exist).[58]

Book printers produced shorter commissions alongside their larger book commitments. The genre of news sheets in German was an early development. Broadsheet publications of other topical issues appeared from printers by the 1480s. But new types of work relating to current affairs would also soon appear, such as a broadsheet telling of Amerigo Vespucci's discoveries in the New World accompanied by a longer publication of his transformative travels: a work of exploration which became another bestseller (Figure 28).[59] In such broadsheets, illustrations were an important part; the non-literate could be drawn to the sheet and secure a literate reader to expand on the significance of the picture. The impacts of printing were numerous – and many would follow quickly after the spread of the new technology and new commerce in Europe.

The Impacts and Context of the Printed Book

The invention of book printing is widely cited as a key part of the developments that heralded the Renaissance, inspired the Reformation and marked the beginning of the early modern era. Yet Gutenberg was a man of the medieval world. Born in or soon after 1400, his was the world of rival princelings, local wars, influential city guilds, ambitious archbishops and a ubiquitous Roman Catholic Church. Only fifty years before his birth, the Jews of his home town Mainz had been blamed for the Black Death and massacred. In 1415, when Gutenberg was a teenager, Jan Hus was burned at the stake for heresy; in 1431, it would be Joan of Arc who met her death by fire during the continuing conflict between France and England, the 'Hundred Years War' which ended only in the 1450s. The Ottoman Turkish conquest of Constantinople terminated Byzantine rule in 1453, and a Muslim emirate controlled the southernmost part of Spain until 1492.

And ironically, printing – the invention which many see as the instrument by which the Protestant movement gained widespread traction in Europe – was established through the commercial advantage it gave in serving the Catholic Church. But the rapid development of printing, as numerous businesses were established in the towns of Europe, had many, varied and profound early impacts. The influence of the printed book on the later centuries is obvious, as the printed word inspired revolutions in society, politics, science, culture and ideas – but the innovation had many effects in its earliest decades, and these effects were widespread across the countries of Europe.

The Book Itself

Among the impacts of printing were those that affected the nature of the book itself. One disadvantage of scribal copying was the risk – indeed the likelihood – of errors. These caused particular concern in the Catholic Church, which required theologically valid and universally similar versions of its religious texts. The missal (the reference source for the mass) needed a reliable and identical version of the approved form and words, and this was achieved only with printing. The importance to the Church of a standardised missal, by which every parish priest was following the same authorised order and words of the mass, cannot be underestimated in the history of early printing.

Before printing, books relied on the accuracy and attention of the copyist; and copyists' errors would accumulate as scribes made their copies not from some master original but from the work of a recent predecessor. Monks might be expected to pay attention to the word of the Holy Bible, but even there, given the length of the Latin Vulgate text, errors could readily creep in. It is hard to imagine much enthusiasm and attention to detail among those required to copy out again and again the pages of the Latin textbook of Donatus – a task sounding more like a pupil's punishment.

The great advantage of printing was consistency and (potentially) accuracy, appreciated by readers of both religious and secular texts. This was important as humanist engagement with the writings of classical authors and the demand for works of science, medicine and law increased. A proof would be taken of each page and checked carefully, with any errors corrected before the printing of copies which thereafter were identical in that edition. This presented a strong additional appeal when competing with the sale of handwritten copies. A later edition could include further corrections to improve accuracy of the text, and the author of a new work could revise their text before another edition. A 'definitive' version of an early text could develop; incompatible versions could be reduced over time.

Such an approach had especial value and immediate appeal recognised by many in the Catholic Church. The future Pope Pius II wrote

The Impacts and Context of the Printed Book 159

enthusiastically in 1455 that he had witnessed the new technology of printing, with bibles 'absolutely free of error and printed with extreme elegance and accuracy'.[60]

In the later era of European manuscript books, some innovations had developed to help the reader, but these were limited by the nature of book creation. The book could not begin with page references because these were as yet unknown. However, side comments (glosses) by religious copyists could be added to a text. With printing, once a complete book had been typeset, it was easy to create a table of contents at the beginning of the book because the pagination was already known before the book was printed. It was possible to include cross references between numbered pages, relevant to have footnotes and stylish to add running heads on pages to label sections of the book (although those had been used in some earlier books). Indexes with page references could now become more common. Once printed sheets had become larger and an imposition established to fold and cut and make the right sequence of pages, numbers or letters would be added to ensure the sections were bound together in the right order.

One feature which could not be achieved by manuscripts was the integration into typesetting of the justified right-hand margin, which added beauty to a book from the beginning of Gutenberg's work. A scribe beginning a line could not predict where the line would end and readjust the text already written. But a typesetter could insert spacing between words to ensure each line ended in the same position. (Initially a hyphen for a word broken across lines lay outside the justification, an approach which was soon changed.)

The size of the printed page was influenced by use: different formats suited different markets. A large-format work would suit use on a library table. A large lectern Bible for a Church would be very different from a book of popular tales or a Book of Hours for personal devotion. Small-sized books were appropriate as student texts to fit into a satchel or backpack. Hand-held books for the individual home reader would be different from those needed in official contexts; and the broadsheet pamphlet had its own requirements.[61] Actual page size reflected different printing machines that were constructed as well as different paper mills, although the terminology such as folio, quarto and octavo was in use to reflect the relation of the final folded and printed page to the full paper sheet.

The potential for accuracy provided by printing was not always reached: poor editing, and the rush to get a new book into the market, did produce books with errors. And once a version of a work by a classical author or a Church Father had been put in print and considered the authentic version, that perception could be long maintained, making it difficult to substitute a more accurate version. Humanist writers, whose scholarship generally benefited from the new invention, would nevertheless draw attention to the inclusion then persistence of errors in the printed texts of classical authors.

The production of the Latin Vulgate version of the Bible in the multiple copies of print encouraged the assumption that the text of Gutenberg's edition was accurate and definitive. Biblical scholarship would face challenges in its subsequence search for authenticity.[62]

The rapid growth of the printing industry suggests the rapid growth of new publications. But it was not quite so simple. Printing was an investment for profit. Because printing required the sale of multiple copies of a work, it was safest to issue books which had already established a significant known market in manuscript form, or where a book had been profitable for a rival printer. In time the publication of safe 'backlist' titles would help fund new works in whose potential the printer had confidence. This contrasted with manuscript production, where the cost of reproducing a book was the same for an established classic or the work of an unknown new writer. In that sense, printing may initially have limited rather than advanced new work.[63]

The rate of survival to today of a copy of an edition printed from before 1500 is, however, high, compared to manuscript titles. A new work created in manuscript might be copied if it had demand, but the copies could readily vanish over time, whereas a printed book would be in an edition of multiple copies – whether 200 or 1,200 – increasing the opportunity to enter long-lasting library collections. But such survival depended both on the importance of a printed work and the enthusiasm and commitment of a collector or librarian; many printed works had little long-term significance and have no examples surviving today.

A 14th-century biblical commentator, or the author of a romance or a medical writer was probably no less talented than their counterpart in the 16th century, but the chances of the later writers remaining for us to see and hold today is far greater. Printing affected not just existing markets but the potential for new authorship and impacted our record of writers before and after the invention of the new technology.

The New Occupations

The industry that began with Gutenberg in the 1450s created, or strengthened, different categories of occupations in the economic world. There

The Impacts and Context of the Printed Book   161

were always readers, librarians, teachers, students and authors. The role of the monastic scriptoria faded, as did the emerging commercial manuscript copyists. Woodblock cutters, illustrators, rubricators and book binders continued for some time. The owners of printing businesses needed investors for capital, specialised staff in type foundry, typesetting and machining; and they needed means by which their books could be promoted and sold as widely as possible.

Printers were the publishers. Printed advertisements began to be used before 1500 but were still rare.[64] Some were for a single book, some to advertise a range of products from a single printery. Those longer lists would meet the ambitions of most printers to sell their work widely, available in multiple cities. Soon too travelling sales agents would represent the printer/publisher, securing orders for the books already printed or due to appear.[65] Freight agents were then part of the business network to move books to the resellers in other towns.

Bookshops had developed before the emergence of printing, to hold manuscript copies for sale awaiting a customer. Florence in particular was a centre of such commercial initiatives.[66] The rapid development of European universities (*studia generalia*) – thirty by 1400, with thirty-four new ones emerging in the 15th century – was a particular stimulus for bookshop development, accelerated by the printing revolution. *Libraires* had already existed as men who would sell on commission manuscript books which had been created in the hope of a market; these same people could readily adapt to the larger market for printed books.

Naturally enough, specialists in second-hand books also emerged: collectors' libraries were dispersed on their death as part of their estate, as shown in auctions recorded in the 1480s which included printed books.[67] Students would dispose of textbooks they no longer needed, and a market emerged which brought no benefit to those who had invested their capital in printing.

Alongside those who established formal bookshops or sold from their printery, hawkers could sell around the streets and could extend the availability of books into rural areas where a bookshop would be unviable. This proved especially suitable for the smaller popular works such as calendars, almanacs, items of romantic literature or shorter works of religious piety. Such vendors also had value in the distribution of religious tracts considered heretical by the authorities; where there was no physical bookshop, it could not be shut down.[68]

Those who had earned their living from manuscripts could adapt to changing needs, or their roles could slowly vanish. Rubricators who created the coloured headings in books were still employed in early books to present a product of manuscript quality, and woodcut artists continued in an important role. Up to 1470, European guilds required any book containing religious images to be made by hand, not by printing.[69]

Perhaps the least financial benefit (as today) was to most authors. To be 'a writer' after the invention of the printed book was very different from the era of hand copying. The printer might consider a right to print the work of a contemporary author as readily as that of a long dead author; a book successful with one printer might be reprinted freely in other cities and countries. The concept of legal copyright would wait another two and a half centuries or more. An author might be supported by a sponsor or might arrange in advance of print a number of sales from which they would benefit. To some authors, promoting their religious or social views, widespread reprinting was more important than the control of editions. It was not until the 17th century that authorship might be considered a source of income; and outright sales of rights to an entrepreneur (printer, bookseller/publisher) rather than royalties on sales remained common, though an additional sum might be offered for a reprint from the same source. John Milton received the sum of just five pounds for *Paradise Lost* (1667), and the same on reprinting. The printer was initially the sponsor of books and authors; the printer's political, religious or cultural values could determine what appeared in print, in a way that would be later controlled by booksellers and eventually by the separate category of publisher.

The Reading Classes

Printing allowed readier access to books for a wider range of readers than manuscript copies; and in turn, this stimulated the gradual development of reading for pleasure among what we could call the emerging literate middle class.

Manuscript works had suited specific markets. The Church required texts (bibles, missals and the like) to fit the needs of the priesthood; and monastic libraries served Church scholarship. In fact, monastic libraries continued to hold a large proportion of their collection as manuscripts well after the printing revolution. Universities' libraries provided resources required by their teachers and students. The wealthy aristocracy might furnish their large homes with manuscript works, for show or for piety. But broader personal ownership of manuscript copies of vernacular literature, or even Latin works of classical authors and later writers, was not readily affordable.

The Impacts and Context of the Printed Book

For the early printing industry, the market for Christian religious books remained strong. It is estimated that of 66 million people in Europe in 1500, some 5 per cent were members of religious institutions – serving in monasteries, convents or parishes.[70] Another estimate that the literate population was about 5 per cent represents a large overlap with these. The sale of bibles remained large, although many bibles in finer editions, as with manuscript bibles, were bought not to be read but to be presented as a pious donation to a library or institution. However, bibles for study did develop, with annotations and study apparatus, and the Protestant revolution reinforced a value of individual attention to the biblical word.[71]

Once the printing industry had developed it provided economies of scale which allowed a book to be printed in as many copies as the market would bear; including ownership by many whose access to books may have ended with their university studies. A measure of book ownership is demonstrated by statistics from Valencia for the period 1474–1550. This suggests books were owned by 9 of 10 in ecclesiastic professions, merchants and 1 in 10 manual workers.[72]

Vernacular popular literature had long existed, a feature of France from the 12th century and England in the 14th century.[73] Travelling entertainers including those once described as minstrels, and later as troubadours, had memorised poetry as well as songs to present to an audience. The option existed to listen to a storyteller or poet or to attend a public reading of a romance or improving moral tale. So, literature had long included entertainment. Increased access to the printed words of poetic and prose works in the vernacular reduced the appeal of performers – although they did not disappear, for only some in society could read for themselves. Ballads were printed for sale, now with the possibility of illustration to enhance their appeal, and satirical songs allowed the expression of perspectives which might have caused more direct offence (or found a less ready audience) in polemical prose (Figure 29).[74] In the 16th century, the market for romances was significant. But it should be emphasised that printing did not put an end to the tradition of oral literature. Only a minority of the population of Europe was literate in the 15th, 16th and even 17th centuries; songs and stories (and of course sermons) remained verbal to access the non-literate majority.[75] Estimates suggest 3–4 per cent of those in German-speaking states could read at the time of the Reformation.[76]

The wealthy with their private libraries may well have indulged in recreational reading. With printing, the reach of vernacular literature was far greater. This meant that reading for pleasure could develop, rather than just reading for the purpose of educational study or religious piety.

FIGURE 29 A *Gest of Robyn Hode*: ballad from ca. 1510 (National Library of Scotland).

Individuals could read to themselves or their family in their own homes, not just listen to a reader performing a work. The literate person could entertain themselves with their own copy of a book in their own home.

However, perhaps more negatively, the spread of printed books, and the reduction of reliance on the spoken word of the religious priest and travelling bard, drew a harsher line between the literate and the non-literate. Those who could read had access to an ever-increasing range of

The Impacts and Context of the Printed Book 165

information and ideas, leaving the majority of non-readers further from the rising middle classes.

The development of printing also increased the distance between urban and rural life. When the book was a missal or manuscript copy of the Bible used in all churches, urban and rural communities had the same relationship to the text. With the printing works based in larger towns, purchasers of books were more local and mainly urban, as were those reached by the new phenomenon of printed pamphlets.

Another feature of the newly developing reading public was the spread for each title. In the past, a sermon from the pulpit of a major church, a ruler's address from his palace or an official's proclamation was heard by a finite audience, who would pass on its content verbally with any flaws that memory created. The words of a printed edition were identical whether read in Amsterdam or London, Paris or Venice. A legal text was available across the whole region to which it applied. Travellers' accounts and atlases could be studied in identical form within different and competing states.

A revolutionary work such as that by Nicolaus Copernicus, *De revolutionibus orbium coelestium*, published posthumously in Nuremberg in 1543, might have been passed from hand to hand in copies in an earlier era despite the opposition of the Church, but in printed form it reached the relevant audience quickly, though with only a small print run.

There had always been religious reformers and political dissidents. Some had attracted local support; some had a message that eventually spread more widely, by the power of voice, but many local heretics could quickly be suppressed by the authorities. In the case of the new preachers of the Protestant movement, the speed by which a new message could be spread in print was a substantial multiple of that of the days of oral or manuscript transmission.

Printing also involved a new approach to the readership and the market. The manuscript book was typically created for a known source, whether a library or wealthy individual, or to lend to a student. Risk was minimal, although the booksellers operating near a university might choose to hold items in stock and order a replacement upon a sale. Printing, by contrast, both allowed and *required* the production of multiple copies simultaneously, even if the printer in the industry's early decades left it to the customer to commission the binding of sheets. A printer therefore had to judge whether the potential market for their work required a print run of 200 or 1,000: this was the range for the first fifty years of book printing, with some editions even smaller and only a few larger in early years of the industry. Therefore, we should not exaggerate the size of the book-buying market in the early period of the printed book – though by the year 1500, print runs numbering 1,500 were not unknown.[77]

The expense of a book lay less in the labour costs of typesetting, printing and binding, and more in the printing premises, the capital costs of machinery and type and especially the high unit costs of paper or vellum. Other than the paper costs, there were of course some economies of scale in a larger print run once the forms had been set up, but not if these could not readily find an adequate number of customers. Risky editions would therefore be balanced, as now, by safer print jobs – textbooks, works ordered by the Church, or jobbing work like pamphlets. The printer had to be a publisher, bookseller and marketer. They had to make a judgement about the language of their books; Latin would reach an international market educated in that language; the vernacular would reach a larger local market but not extend across borders (although migrations that followed religious upheavals might make those markets more fluid).

Language and Identity

The printed book helped to set the languages of writing and education in permanent form. The many spoken languages of later medieval and early modern Europe (as elsewhere) varied substantially in regional dialect, by class and by style. While local languages and variant patterns of the spoken tongue would remain, printing encouraged standardisation in the written word. The author and the printer (as publisher), seeking to reach as large an audience of readers as possible, would use a language for a book that was typically close to those used in official documents of church and state.

Books in print continued to be in a medieval Latin that was relatively standardised, alongside editions of the classical authors of ancient Greece and the Roman world. Latin remained the international language (such as in works in philosophy or science) as well as the language of the Roman Catholic Church. The great advantage of printing books in Latin was that they could sell across state borders. Only as the market in individual local European languages grew did Latin reduce as a proportion of books. In the period of 1450–1500, it is estimated that 77 per cent of books were in Latin.[78] But then it changed. One estimate of books printed before 1600 noted an even division between 166,000 works in the scholarly and religious languages of Latin and Greek, and 180,000 in local vernacular languages which would have a broader but more localised audience.[79] And to the end of the 16th century, only 10 per cent of books printed in England were not in the English language, a stark contrast to the mix of vernacular and classical languages of Continental printing.

Incunabula brought into broader currency works in German, Italian, French, Spanish and English, all of which had wide variety in their locally spoken forms. The impact when printed books spread to other countries was at least as great, and printing set a standard in Hebrew.

So whereas communities in the Italian peninsula might speak in a range of linguistic styles, the publication, distribution and sale of important and appealing works in a literary version of Italian could establish a linguistic norm which would influence education and therefore the written style of more routine uses. The *Divina Commedia* of Dante Alighieri was printed in an edition in 1472 in the Italian town of Foligno by German Johann Numeister, and other editions appeared in Florence and Venice. Reflecting the Tuscan language of its author in the early 14th century, its influence helped develop this as the recognised Italian language for authors, even though pronunciation continued with regional forms. As one history of the book suggests, printing 'celebrated and enlarged the capacities of the common tongue'.[80]

The spread of Protestant literature in the vernacular – such as Luther's German translation of the Bible and his other writings – could inspire in these languages a reverence previously associated with the Latin of the Church and Vulgate Bible. A greater standardisation of written German gradually developed, in contrast to the many regional forms of spoken German.

Printing in England in the vernacular for a local market helped to establish a standard English, which can be traced back to the works of William Caxton and his successors from the 1470s. Although Caxton published works first written in English, such as Chaucer's *Canterbury Tales*, it was new translations into English which helped establish a standardised literary form of the language which became that used in education alongside Latin. An official form of the language, 'Chancery English', was in use by the 15th century, and since Caxton needed to reach as wide an audience of educated literate and motivated readers as he could, he adopted for translations that version. However, his typesetters (presumably working with a counterpart dictating the text) varied in their spelling, and indeed it would take until the mid-17th century for a more standardised spelling of English to be fully established.

Note too that the spread of vernacular works had further-reaching impacts. As the reading classes grew and shared a common form of a language, so too came the possibility of a greater unity and identity around that language. The movement away from Latin – the language of religion and education and ideas shared across much of Europe – can perhaps be linked to a rise of national pride and national identity.[81] Those who read, were educated in and wrote in, for example, Italian or German shared a common identity that cut across the boundaries of states and principalities. Thus, the commercial necessity of reaching as large a market as possible beyond that for Latin works helped forge and unify languages then used in education, which in turn strengthened the role of language in defining the identity of those who wrote and read in it.

Genres

European manuscript books had covered a range of genres, expanding in the later medieval centuries with the growth of universities and with the increase in vernacular texts, especially of popular literature. Printing allowed certain categories to develop faster. Of incunabula – books of the first fifty years of printing – 45 per cent were religious books, 30 per cent literature, 10 per cent law and 10 per cent what we could broadly group as 'science'.[82] In the growth of printing through the next century, the proportions would change dramatically as the latter categories grew.

With printing, the opportunity to supply (and in turn stimulate) a growing and diverse demand for books could be met, and the incentive for new writing became correspondingly greater. Devotional books remained a good market, and works on philosophical and social topics increased in reflection of (or reaction to) humanist developments. Creative works of romantic literature for entertainment and leisure went alongside works of history and tales of travel, while books of personal guidance and moral education, designed by the old for the young, were an additional genre. Printing might have begun with the demand created by the manuscript book, but it could generate its own new markets.

Manuscript books whose contents required illustration, such as books of medicine and science, required far more work than books with text and decorative lettering, and therefore a higher investment.

The evolution of the printed book allowed a greater level of illustration and greater focus on accuracy and consistency than any hand-drawn additions to manuscripts. Manuscripts from the earlier 15th century had included woodblock illustrations for such scientific and medical works, and printed books could incorporate a page of illustration cut on a woodblock (or for fine-line work engraved on a copper plate). Before long, the ability to create and print books with sometime lavish levels of illustration gave the genre a major boost. In the later 16th century, the engraved metal plate became the norm.

Information on natural history and geography as well as science and medicine became available in standard editions, where a wide audience were looking at identical images – though it was a different question whether these were accurate and (in the case of atlases) up to date with the latest knowledge. The option to add new material to works of scientific reference was an incentive for new editions by new printers. But not all science was for the serious specialist reader. A natural phenomenon such as a comet or a biological rarity such as conjoined twins could provide the subject of an inexpensive illustrated pamphlet for sale in the public marketplace.

The changed world revealed by the explorations of Christopher Columbus, Amerigo Vespucci, Vasco da Gama, Bartolomeu Dias and others led to a burst of travel publications of variable reliability. A letter from Christopher Columbus reporting his journey in the New World was a short pamphlet that sold well; another work by Amerigo Vespucci proved a best-selling item from 1504 onwards.[83] The genre of travellers' tales developed. Sebastian Münster's *Cosmographia*, issued first in 1544, was a remarkable account of the world's geography in 640 pages, 520 woodcuts and 24 more maps, and even longer in subsequent editions.

The first printed maps from the 1470s onwards reproduced manuscript cartography and traditional world images. Piety could typically maintain the image of the Holy Land at the centre of the world. But printed maps, and books presenting maps, were a major genre of the 16th century, reflecting both the politics of European states and the revelations of the Age of Discovery.[84] At scale, they began to serve the needs of navigators.

Laws and legal documents had been the subject of written records for some time, but manuscripts of laws and statutes were available only to those who could access court collections or legal libraries. Now, it was possible for the authorities to publish and distribute new statutes and accessible volumes of collected laws; and for printers to produce law textbooks for students. Courts would hold identical copies of reference works, and lawyers could have greater access to personal collections of such legal reference works. Indeed, already in the 15th and early 16th century, booksellers could be found in and near the Paris law courts. Not only lawyers but merchants and others might require reference to the law. And while books on religious or political matters might fall foul of changing affiliations, and literary publications would be subject to changes in popular tastes, the reprinting and updating of books of the law could provide a safe and reliable income for printers.

Another genre to benefit from printing was music. The copying of musical notation by hand was a far more taxing job for scriptoria than text; the main demand had been for purposes of religious chant. And of course not every new musical composition or arrangement would secure a potential market large enough for a printer to sponsor, so hand copying would remain. But printing sheets of music for sale began as early as 1473 and would benefit from the adoption of a standardised notation. Initially, there would be double printing of each page: first, the aligned stave lines, then the notes overprinted on these, a demanding task to secure accurate alignment. In the 16th century, some 2 million copies were printed of hymns and song sheets, with an estimated 6,000 editions of music publications between 1520 and 1600.[85]

Education

The burst of education in Europe in the 15th and 16th centuries stimulated the production of printed books, which in turn rewarded educational institutions and their libraries, teachers and students with new tools to expand learning.

With the 30 universities at the start of the century more than doubling in 100 years, the market significantly expanded for library works of reference and for student textbooks. The entrepreneurial booksellers who served the university communities with the sale of manuscript books were now establishing new relations with entrepreneurial printers.

Patterns of teaching and patterns of learning, and the relationship between students, teachers and their libraries, could change with the printing revolution. In the medieval university, it was common for texts to be dictated by the teachers and copied down by students as best they could. Students might be expected to memorise or make their own notes on the topics of a lecture. Sometimes, teachers would take advantage of the new technology by publishing their lectures, in some cases to prevent others doing so from notes taken during their delivery.[86]

The resources of a university library were important, if limited for the large numbers of users. The development of manuscript copies of a text to rent out part by part for copying – the *peciae* – was an initiative of commercial booksellers in Paris.[87]

Manuscript copies had frequently been corrupt, not least as students undertook to make their own copies from others' work, without necessarily the highest confidence in the Latin used. Glosses and commentaries in library manuscript books were added to help in study but might just as often confound.[88]

Print provided students with accessible and relatively reliable versions of the classic texts being studied. Access to classic works no longer required visits to libraries or dictation from university teachers. A printed work could be bought new or second-hand, or still hired in parts to be copied by an impecunious student. A printed edition in libraries or purchased from a bookseller could be relied upon to have an identical text without a variation of errors: teachers and all their students could be working from the same version. Whereas a scholar, teacher or student might have been reliant on old manuscript books that formed the core of a university library, new books added to library collections and new ideas within them could take precedence.

To secure work needed for their institution, in 1470 teachers at the Sorbonne in Paris took the initiative to commission and provided facilities for printers.[89] Yet it was also Sorbonne academics – those in the faculty of theology – who would later take a lead in the censorship and oppression of heterodox books and their originators. As late at 1526, Sorbonne theologians moved against the use of the French language for scripture, and French bibles found their printers in Antwerp and Geneva, though also in Lyon.[90]

The study of subject areas such as medicine, with the importance of illustration rather than just the words of classic texts, was transformed by print. Teaching no longer relied just on the verbal and reference to manuscripts. Of course, medieval medical manuscripts were illustrated, but increasingly printed books could include a range of illustrations which would provide a reference source not just for students and their teachers but for those who proceeded to professional work in medicine and other developing areas of science.

Writers and printers turned to creating new 'textbooks', accounts of a study subject which could be sold to students who would study from identical copies approved by their teachers (Figure 30). As the scope of university teaching developed, such a development would bring benefits to all parties. Those who proselytised the pedagogical value of the new type of product could boast their efficiency: 'Latin in eight months, Greek in twenty days, astronomy in eight or ten days, philosophy in a month or less.'[91] The great humanist scholar Erasmus would write textbooks for student use alongside works of polemic and interpretations for a wider audience. The French Protestant writer Petrus Ramus (1515–1572) would author over fifty books reflecting his systematic approach to the nature of

FIGURE 30 Textbooks in use: a woodcut from Perotti, *Rudimenta grammatices*, 1495 (photo credit: Album/Alamy Stock Photo).

teaching, achieving major sales for the printers involved. The first centuries of printing, which made knowledge accessible to a far wider range of readers, stimulated 'the creation of a vast range of treatises that tried to impose and artistic and coherent form on virtually every aspect of human conduct'.[92]

The Impacts and Context of the Printed Book

Schools and colleges to educate the young before the stage of a university grew in numbers during the 16th century (like my own London high school, founded in 1561, with a committedly Protestant head teaching English grammar alongside the classical languages). The use of printed textbooks enabled a transformation in teaching emphasis and style.

Religion had been a dominant part of pre-university education in medieval Europe. It would continue to be a requirement in the syllabus of schools, dependent on the prevailing religious orthodoxies. Direct study of the biblical text itself, central to Protestant teaching, became a feature in English grammar schools only from the 17th century; in the reign of Catholic Queen Mary (1553–1558), access to the Bible had been inhibited.[93]

The increase of scientific and other new directions of inquiry in the Renaissance was not reflected in school teaching.[94] The emphasis lay in grammar, which took priority over subjects such as rhetoric, dialectic and logic which had been part of earlier elite education. Grammar meant primarily Latin, and the ability to translate, to write and to speak Latin was a priority. Key Roman authors were read to establish style as well as to influence by their contents and the values they imparted. The educational fashion towards selecting for study the best Latin authors, already noted earlier in the 15th century, was accelerated by the availability to teachers of reliable printed editions. There were 300 editions of Cicero printed in the 15th century, and 1,500 in the 16th: one estimate is of 2 million copies for students.[95] Although dictation and copying were important, a class of school students could work from the same texts and learn from the same printed grammars, phrase books and collections of selected passages.[96] Greek began to be added to the syllabus in many schools, and Greek grammar textbooks were available for English school students from 1530.[97]

But while the emphasis on classical languages, and Latin grammar in particular, was a feature of the schools' educational syllabus in England and elsewhere, this does not imply that the breadth of learning of humanist scholarship and its questioning intellect was the norm in the formal education of young men of the 15th and 16th centuries.

Intellectual Enquiry, Renaissance Humanism and Scientific Development

The spread of printing stimulated intellectual endeavour of every kind. It made widely available books of classical authors whose work was the focus of the developing humanism. It made knowledge available outside the universities (where Latin retained its role) because books were now widely available in the vernacular. And it served as a catalyst to transform science from an area of theory based largely on classical authors to a practical endeavour of transcontinental research and development.

The movement we call humanism as a major feature of the European Renaissance was notably an emphasis on recovering, exploring and developing the ideas and literatures of the ancient Greek and Roman world. It was not in essence a challenge to religion and religious authorities, but a means of leapfrogging the inward-looking 'scholasticism' of medieval Christendom and the current issues and perspectives of the Church. There were indeed major innovative thinkers and writers such as Rodolphus Agricola or Desiderius Erasmus, but a move of education from religious scholasticism to the languages and select authors of the Greek and Roman world could bring its own narrowness. A shift of orthodoxies in education could suit changes in the world of European politics. As a critical account of those educated in the early stages of the new approaches noted, 'fluent and docile young noblemen were a commodity of which the oligarchs and tyrants of late fifteenth-century Italy could not fail to appreciate'.[98]

Printing contributed significantly to the humanist movement in making readily available editions of what had previously been limited mainly to the wealthy and to libraries. Enthusiasm for the book was key to humanist outlooks. The writings of ancient Greece and Rome could bring old perspectives to the new generations, whether from literature, history, philosophy or scientific theory and description, or indeed the early Church Fathers. They were supplemented already by the 1470s with the printing of texts derived from within the Islamic world such as Persian scientist Avicenna (Ibn Sina) and Jewish philosopher Maimonides (ben Maimon). And after the Ottoman conquest of Constantinople in 1453, coinciding with the birth of European printing, Byzantine scholars fled west, bringing with them manuscripts of Greek authors to add to the canon.

Printing stimulated the use of Greek works and the educational study of the Greek language in western and central Europe. While religious readers and students could readily build on direct knowledge of the Greek New Testament (with Protestant leaders especially favouring new translations and interpretations), the classical Greek authors played a key role in the humanist-influenced educational changes. Therefore, the printing of textbooks to teach students Greek developed as a profitable venture.[99]

The Impacts and Context of the Printed Book

Humanist teachers could be based in universities or remain within the Church (or indeed, both); but the attention drawn to the ancient (and by definition pre-Christian) writers could lead students and other followers directly to printed copies of the relevant book. The original words could be considered in newly standardised editions seeking to use where possible the original Latin or Greek, without the overlay of medieval Christian commentators, although new commentary inspired by the humanist movement could be added. Once set in type, carefully proofread and printed in multiple copies, the original text of a classical author could be considered more reliable that the single copy generated by a weary copyist in a scriptorium. Nevertheless, editions of classical authors represented only one in twenty editions from the first fifty years of printing.[100]

A Greek fount developed for Venetian printer Aldus Manutius was used for editions of Greek classical authors; the same printer developed an italic type for Latin. Greek typesetting had first been used for didactic grammars and dictionaries. By the early 16th century, founts for Hebrew script were in use. Books in Greek and Hebrew were at first too expensive or too limited in their distribution to suit the growing educational demands.[101] But as Protestant reformers put emphasis on biblical texts and new translations into the vernacular from authoritative versions of Old and New Testaments, Greek and Hebrew printing followed Latin in establishing profitable type and typesetting.

The leading Dutch humanist Desiderius Erasmus (1466–1536) was born just as the first printing in the Netherlands began. He therefore developed his writing in a context of print publication, with his edition of the New Testament in Greek issued in 1516 (the Latin Vulgate being the version used by the Church). In addition to his works on religion and philosophy, Erasmus wrote textbooks for educational use. His own writing (often initially in Latin then in translated versions) would make him one of the best-selling authors of his time – one estimate was that in the 1530s 10–20 per cent of all book sales were authored by him, with a million copies of his books printed in his lifetime and 2,500 editions through the 16th century.[102] He specified the invention of printing as a liberator from a world where 'those who wielded the sceptre held forth on the least informed teachings with never a murmur of discontent'.[103] Yet he was not averse to seeking to suppress the printing of works that criticised him.

The availability of identical books across national borders meant that knowledge and ideas could quickly spread widely, stimulating new undertakings in state policy as well as in commerce and intellectual endeavour. It was said that via the printed word the navigator and explorer Columbus learned that the Earth was round and that therefore a journey west should reach the east – the Indies. Further conquests in the New World had in mind the adventurers' tales that had appeared in print.[104]

With illustrations, printed books could spread more than new ideas. Ideas in architecture were dispersed through the agency of print; a public or private client could indicate to their architect the elements they favoured in a new commission. Andrea Palladio's *Four Books of Architecture* published in Venice in 1570 was particularly influential. An edition of the architectural text by Roman author Vitruvius had appeared as early as 1486–1487. Even fashions in costume or decoration could been influenced by those in the illustrated books.

Not least, printing had a transformative role in science, technology and medicine. The classical works might have afforded respect to the insights of great authors from the ancient world, and more recent authors from medieval Islamic and Christian states, but now scientists could see work freed from the constraints of manuscript and Church, to be published as a commercial venture by printers who anticipated an audience across Europe. Science involves communication and the exchange of ideas. With print, these ideas need not just be communicated with others by correspondence or in person, but in publicly available statements.

Scientific writings were not the natural or easiest tasks for medieval copyists. Printing technology increasingly allowed scientific illustration to complement text, transforming or even creating new genres. With printing, once the sometimes-complex text of a scientific work had been set up, and combined with the required illustrations, all saleable copies could be produced to meet the expanding market.

Inevitably, some of the scientific works of classical authors were first to be printed, before European writing was added to the inventory of print. In the first half-century of print, books on scientific subjects were typically unillustrated, reflecting the respect still paid to the established authors. There were twenty editions of the works of Arab medical writer Avicenna (Ibn Sina) before 1500. But thereafter, new writers and thinkers in science and technology, in anatomy or physiology or medicine, could hold before them copies of the illustrated work of others as they performed their own experiments and wrote their own books for print. Medical professionals could have information set out where they worked. Andreas Vesalius' *Anatomical Tables* appeared in 1537–1538.

Printed books became important tools of research and new thinking. Works of theory could now be read by those involved in practical work. Innovations in technique and application could be communicated readily to other technicians and practitioners across the continent, as well as to those whose interests were primarily scholarly in nature and whose knowledge was now expanded by the practical experience and experimentation of others. The production and distribution of printed books brought about the interaction of theory and practice – of those involved in developing and refining ideas and those involved in testing and developing the practical application of ideas.

The revolution that made possible the creation and distribution of multiple identical copies of a work – the printing revolution – was a catalyst for the driving forces of Renaissance humanism, new exploration and new popular ideas which in turn would be disseminated by print. In those societies open to new directions of knowledge, ideas and experimentation, it could break down barriers between scholars with interests in scientific themes, and those with daily lives in practical medicine, engineering and technological development. And as a contributor to such developments, the wider world revealed to Europe by the first century of printing would challenge and change the world view asserted by biblical tradition.[105]

Ignorance

Just as the modern Internet (websites and social media) has allowed the rapid spread of conspiracy theories, New Age cults, fake news and wild and wacky eccentricities, so the development of printing allowed the spread not just of knowledge, literature and carefully argued theological ideas, but also alternatives. With the rapid development of printers across Europe, competition meant that they sought works for which they could obtain a ready market. Falsehoods could spread as rapidly as truths. As science developed, so pseudo-science, a 'natural history of nonsense', also developed.

The definition of what is valid, tested and true, and what is fantasy and invention, is subjective and changing. Popular taste, itself subject to rapid change in the early modern era, could determine what would sell and be read. And scientific errors in a widely disseminated printed and reprinted edition could appear to have a level of veracity that would outlast new inquiry.

As well as books on scientific topics we would recognise today, works on astrology had an equal reputation and audience as valid science and a respected area of study alongside astronomy and cosmology in the period when printing developed. The annual almanacs from astrologer Nostradamus published from 1552 were major selling items containing predictions, and which have developed a following through to modern times.

The power, authority or influence of Church and state might limit a printer's initiatives although (given the cross-national pattern of the early book market) the prohibition of a book by a Church or state leader could be cited to publicise an edition elsewhere.

Mysticisms of various kinds could take advantage of the technology – or rather printers could, ever seeking new markets. Many of the most popular religious works issued by early printers were works by Christian authors writing in a tradition of Christian mysticism. But beyond this, printing allowed the re-emergence of Hermeticism, with the publication of works translated from Greek and Roman sources. These presented mystical beliefs associated with a supposed Hermes Trismegistus, seen as reflecting ancient Egyptian wisdom.[106] *De potestate et sapientia Dei*, translated by Marsilio Ficino, was issued in Treviso in 1471, followed by other editions. By contrast, some purveyors of mystic insights would emphasise the importance of their mysteries *not* being revealed in print to a wider public.[107]

With no independent sources of review, readers could not distinguish works of fact from works of fiction (assuming they might wish to). Accurate accounts of distant travel sold alongside fanciful works like the 13th-century *Travels of Sir John Mandeville*, of which thirty-eight editions had been printed in different languages by 1500. Mythical national histories had at least as much appeal as those based on more rigour. Print brought every kind of book to a wider audience.

Public Discourse and Propaganda

The flexibility of printing meant that short pamphlets or broadsides (*flugschriften*) in a popular format could be readily produced and distributed. One estimate suggests that 9,000 pamphlets appeared in the first thirty years of the 16th century, but the figure may well have been more. Printers were no longer reliant on the success or failure of their investments in major new books, in print runs which might or might not sell through, but could maintain their cash flow through small commissions encouraged by the rapidly changing world. Jobbing printing had existed alongside book printing from the very beginning of the industry, with the Church's commission of printed indulgences.[108] In the first half-century of print, indulgences made up 30 per cent of the single-sheet printed items, and the majority of the others were commissioned by political or religious authorities. What had previously been proclamations posted publicly in a limited number of

The Impacts and Context of the Printed Book 179

handwritten copies could now be widely distributed, and rulers could boast of their achievements in leaflets where text was augmented by illustration.

While literacy was far from universal in Europe, there were sufficient numbers of people educated to read in the vernacular language of their region that ideas could spread rapidly, as a pamphlet was passed from hand to hand or posted in a public place, where the literate could read out its contents to those around them. The highest levels of literacy were found in the cities: perhaps a third of the adult males in 16th-century Venice, a third to a half in London.[109]

The Peasants War of 1524–1525 in Germany was not a war of the literate classes, but illustrated pamphlets contributed to the development of the movement and were used as weapons in that war. Tens of thousands of copies of the *Zwölf Artikel* outlining peasant movement demands were printed, reprinted and distributed. Despite support by some of the Protestant leaders, Martin Luther promptly used a printed pamphlet to distance himself from the peasants' rising.

Political, legal and information pamphlets could help keep printers' cash flow positive, as long as the content did not clash with the authorities. Leaflets might also be poems commenting on current issues (some with illustrations to add to their appeal), and the use of the single printed sheet to distribute or flypost had become relatively common by the end of the 15th century. A complex theological position could be presented more widely in a short pamphlet, reaching an audience that a serious book could not; pamphlets would play an important role in the early development of the Protestant movement.

Reformation: The Religious Revolution

There is widespread acknowledgement that the spread of Protestantism from the first half of the 16th century owed much to the invention of printing and the ability of the new theologians and religious organisers such as Martin Luther (1483–1546), Huldreich Zwingli (1484–1531) and John Calvin (1509–1564) to disperse their ideas and messages more widely through the means of print. Such a resource had not been available to early religious critics such as Englishman John Wycliffe (1320–1384) or Czech Jan Hus (1372–1415), whose innovative ideas could spread only through word of mouth and manuscript copying, important though that was when written in the vernacular rather than just Latin. The Catholic Church's loss of control of publications had been well established before the printing revolution.

The institutions of the Church were responsible for book production in Europe until the emergence of commercial copyists after the 12th century – so they had controlled not just how books were created but what was created. With the development of universities as institutions separate from those of the religious orders, the business of secular booksellers (stationers) emerged to supply the new market of scholars, students and graduates.

Not least for the Protestant reformers was their emphasis on the Bible as the source of religious belief, in editions of vernacular translation rather than its interpretation in the hands of a small elite of priests. While these were accessible for all literate Christians to read in their own copies, costs of purchase remained high, and literacy rates were modest. Therefore selection, reading, presentation and interpretation of biblical and other religious texts would long continue to be dominated by religious leaders in Protestantism as it had in Catholicism.

Clearly, Gutenberg was not related to any Protestant movement, with his workshop printing indulgences for the Catholic Church and issuing his Bible in the official Latin translation of the Vulgate (there would be ninety-four editions of the Latin Bible from printers before 1500). The 1466 German-language Bible is noted for the poverty of its translation. In subsequent editions of the Bible, significant levels of illustration had a theological effect, complementing the imagery present in church buildings.

While Catholic orders such as the Dominicans forbade lay people from reading the Bible, the Augustinians promoted public access and use of biblical texts, and Martin Luther was an Augustinian monk from 1505. The Augustinian order was particularly influential at Erfurt University, where it is possible that Gutenberg himself studied and where Luther became a student from 1501 to 1505.[110]

There had been vernacular translations of all or parts of the Bible into a range of European and Middle Eastern languages before the age of print, though not without controversy.[111] Catholic Church authorities varied in their attitudes to such translations, with greatest concern when a specific translation was associated with a group considered heretical.

In the views of the Protestant reformers, the written word was the word of God, and its authority went beyond (or could replace) that of the priests. Access by individual Christians directly to the words of the Bible, bypassing the religious authorities, was the key to salvation. The new printing industry could trade on the potential for these among a devout or critical reading audience. While the first printed editions in vernacular translation had

The Impacts and Context of the Printed Book

a market among scholars and churchmen already familiar with the Latin Vulgate, those produced for the laity soon came to predominate.

The first French-language printed Bible was produced in 1479; Italian, Dutch, German and Czech language editions soon appeared, translated from the Latin, not the original Hebrew and Greek. The Protestant translator of the Bible into English, William Tyndale (1494–1536), used those original sources in his English-language translation. One frequent form of printed edition was more affordable to a lay readership: just a part of the Bible, whether one or more gospels, an edition of Psalms or Proverbs or another individual book.

Luther's '95 theses' questioning aspects of Church positions and practices on indulgences were in fact written in Latin, first in a letter within the Church in late 1517, and then famously posted in manuscript copies on church doors in Wittenberg. The printing industry and commercial enterprise soon ensured their widespread distribution and sale in pamphlet form, initially in Latin and soon translated into German. Wittenberg itself would develop into a major printing centre as a result of Luther's writings.

Luther's subsequent sermon on the use of indulgences and the importance of God's grace could reach a broader audience through the use of the German language and willing printers. This was followed by his many new writings, both critiques of official Church doctrine and presentations of the new theologies he was developing, which found a new and ready audience in print. The wide distribution of printers throughout Europe ensured that translations of Luther's works rapidly appeared in other languages. If the printing industry was a gift to the development of Luther's Protestantism, Luther and the popularity of his writing provided a gift to the printing and bookselling trades, supplemented by the flow of works criticising him.

Eighteen different German-language bibles had been issued before the time of Martin Luther's public life. Luther's own translation of the Bible into German from 1522 to 1523 appeared alongside his innovative and revolutionary theological preaching. This gave him a market advantage and a reputation of authority to those who looked to his ideas. Luther's Bible translations became strong sellers and profitable ventures for printers. Perhaps one-third of German book printing in the first fifty years of the 16th century was of works from Luther's pen, with half a million copies of his biblical translations completed, making money for printers and booksellers.[112] As with all future translations of the Bible from Hebrew and Greek, the translator's own beliefs and affiliations influenced the phrasing and language used in the vernacular versions.

Luther overtook both Erasmus and the Italian cleric Girolamo Savonarola as the most popular contemporary author in Europe. It is estimated that already by 1520 there were 250,000 copies of his works in print.[113] His *Address to the Christian Nobility* sold 4,000 copies in just a few days. Luther's *Treatise on Good Works* issued in 1520 was reprinted eight times in that year, with six reprints the following year.

Between 1520 and 1525, some sixty authors loyal to the Catholic Church responded with over 200 books and shorter pamphlets against Luther and his beliefs, but aimed at a different audience, being written in Latin rather than vernacular languages.[114]

Luther was not, of course, the only early Protestant author. Zwingli was influenced both by Erasmus and by Luther, developing his views first in sermons to his Zurich church congregations, then from the 1520s in short published pamphlets and longer works including printed editions of sermon, which together represent a substantial corpus.

John Calvin presented his developed theological views in a consciously polemical vein (he had first trained as a lawyer). A formal statement *Institutio Christianae Religionis* was published in Latin in 1536 (and in French five years later). Translations into other European languages soon followed. Reflecting his emphasis on the Bible as a basis for faith, Calvin's commentaries on individual books of the Bible were also influential printed works. Many theological studies would follow. He wrote over 100,000 words a year to appear in print in the last fifteen years of his life.[115] Calvin's emphasis on reading the Bible as a source of salvation also correlated with the role played by the book-printing industry.

The move away from the Roman Catholic Church in England was, of course, initially political more than theological. An injunction of 1536 on local parish churches to acquire and make available the Bible in English was issued despite the absence of an approved translation; William Tyndale's revised edition of the New Testament had been printed in Antwerp in 1534 before his execution in 1536. Religious works in the vernacular had been popular in England, with over 500 different editions of Books of Hours before the 1530s, the Benedictine order being the only printers active outside of London.[116] In the religious conflicts and changes under the Tudors, many works appeared in English denouncing the prevailing orders, issued by authors in exile through Continental printers.[117]

Considering the rapid development of Protestantism, what could have been local debates, and perhaps a slow development of heterodox positions with manuscript copies of relevant documents surreptitiously passing from hand to hand, thus spread with remarkable rapidity because of the printing press. Protestant leaders and writers were not slow to acknowledge the role of printing in spreading their version of 'the Word of God'. In John Foxe's famous and massive *Book of Martyrs*, published in England in 1563, he noted: 'The Lord began his work for His Church not with sword . . . but with printing, writing and reading.' The new religious views may have spread with word of mouth and travelling preachers, but print could go wider and faster throughout Europe.

But in its origins, the printing press had served the established Catholic Church in new ways; if it was a handmaiden to the new Protestant developments, its uses and significance to religion would continue to be more diverse.

Anxiety about Printing

The Counter-Reformation from the mid-16th century reflected in part a response to the challenge to Church authority that had come about through the broader access to printed works, including the spread of vernacular translations of the Bible and other religious texts. Such concerns had begun soon after the potential of the printed book had become clear. Filippo de Strata, a Benedictine friar, protested to the doge of Venice about the danger implied from local printers and urged their control, in a polemic written in 1473–1474. He was concerned for the negative impacts arising from public access to literature: it would be bad for scribes, it would be bad for authors and it would be bad for public morals. Printers 'shamelessly print, at a negligible price, material which may, alas, inflame impressionable youths, while a true writer dies of hunger'.[118]

In 1485, the Archbishop of Mainz had complained about works 'that were written on divine matters, and the cardinal issues of our religion, translated from Latin into German and falling into the hands of common folk, not without bringing dishonour to religion'.[119] In 1487, Pope Innocent VIII threatened excommunication on printers who issued works contrary to the Catholic faith without permission.[120]

Pope Leo X in 1515, just before Martin Luther launched his theological challenges to Church practice, had warned that the reading of printed books could lead readers into errors of faith and errors in their lives and morals and create 'manifold troubles'. The service provided to the Protestant movement by the printing industry proved his prescience.

Inevitably therefore, censorship became a feature that would affect the printing industry. Initially, the authority given by a local ruler to printers working in his domain could determine whether a work was avoided or withdrawn. Resistance to Protestantism by a ruler could produce a firm response, as exhibited by Charles V in the Holy Roman Empire. But enthusiasm for censorship would not be restricted to Catholic orthodoxies; rulers of states whatever their religious affiliation would seek to exercise some control over what was printed in their realm well into the 17th century, if not beyond.[121]

In the second half-century of printing, with the rise of Protestant teaching and alternative ideas not in line with Church orthodoxy, the formal classification and suppression of books grew. Published lists of banned works developed – in Holland in 1529, Venice in 1543, Paris in 1551 and in Rome under the Vatican authorities from the 1550s – some banning all work by named authors, and the Roman Index of 1559 also named banned printers of Protestant works.[122] These editions of an *Index Librorum Prohibitorum* only served to stimulate interest in those works considered unsuitable to read and a market for the printers who continued to produce them, often selling from areas under Protestant influence into those where Catholicism remained dominant.

We must admire the initiative of the Paris booksellers who suggested in 1545 that the books they stocked which were now banned could be sold with a list inside of the passages which were now to be avoided – an idea which did not get authorised! But the censorship that accompanied religious tensions was not usually so amusing: printers and booksellers of forbidden and heretical works were arrested, imprisoned, even executed as the printed word challenged the Church authorities and their state and university supporters.[123] French printer Étienne Dolet was burned at the stake in 1546 together with his books, on grounds of heresy and with the active condemnation of the academic theologians at the Sorbonne, the source of earlier enthusiasm for what printing could offer their students and their libraries.

Burning books as well as burning their authors was symbolically important. Holy Roman Emperor Charles V showed his piety and loyalty to the Pope with his campaign of book burning in Belgium and Holland, and Luther's works were burned publicly in London in 1521. But in similar vein, Luther organised public burning of works by Catholic authorities.[124] He had hoped to include some of the books of early Church authorities in his bonfire, but editions of their work were too precious for their owners to part with, however committed they might have been to the new religious cause.

Conclusion

The early history of the European printing revolution shows how it coincided with major changes in European society – humanism, Protestantism and the Reformation, and the impact of the arts, science, cultural and political changes of the Renaissance. As the later Middle Ages transitioned into the early modern period, the facility of printing accompanied and often facilitated major changes.

Some have suggested that the historical contribution of the printing revolution to social change has been overstated, and these debates will certainly continue.[125] A challenge is to distinguish between impacts associated with the development of the printed books and social changes where printing was a minor or even irrelevant contributor. Much of the debate relates to whether printing *continued* to be an agent of change; it is easier to assert the effects and importance of the introduction and rapid spread of the printing press in Europe in its the first half-century or so. We need always to remember correlation is not causation.

In East Asia, the impact of print had been largely controlled by the authorities, who combined secular power and religious backing. Printing was often an official exercise, not one undertaken by entrepreneurs for profit. The written languages of China, Japan and Korea did not lend themselves to the innovations of movable metal type. In the Islamic Middle East, awareness of Eastern printing did not transform into a local printing industry, and the manuscript retained its position of strength.

It was in Europe that the printing revolution served as and proved a catalyst for social changes. The extraordinarily rapid development of printing enterprises across Europe demonstrated that it could be a profitable enterprise. Printers had to balance products with a known market – indulgences, Books of Hours, short pamphlets or textbooks – with more ambitious if risky book projects, where they took the financial risk on what size of print run would sell. They also had to balance commercial imperatives with the risk of offending secular or religious authorities.

The revolution associated with the name of Gutenberg was all the more remarkable in that it brought together a series of inventive and appropriate techniques almost simultaneously: the creation of moulds from which to cast type; the casting of type on a base which would be shaped to give a common height; the setting of type in formes and formes in pages; and the economies available as type was redistributed back for reuse. Printing needed paper to be damped to the correct levels, type had to be prepared with the right kind and amount of ink, a printing press had to be constructed and used to provide just the right amount of pressure to convey a sharp image to the paper and the same had to be repeated on the other side of each paper sheet before these were brought together into a book. All these arrived together – the first books from Mainz had a style and appearance that strike us still in their perfection.

The spread of the printed word both reflected and encouraged the growth of literacy (and indirectly helped standardise the written languages of Europe). Readers could access the words of their religion, or the words of those proposing challenges to the established Church orthodoxy. Writers could present to a wide audience their ideas in science, medicine and philosophy; works of literature, tales of adventure and maps of the expanding known world. Students could access identical versions of classic texts, and the new category of textbook developed for sale. Print could produce pamphlets in their thousands to influence or inform. The speed of such changes meant the world would not be the same again. In communication – and in all those social effects where communication plays such a central role – printing can be compared back to the invention of writing, but also forward to the harnessing of radio waves four and a half centuries later.

CHAPTER 6

Communicating Wirelessly

> *When wireless is perfectly applied the whole earth will be converted into a huge brain, which in fact it is, all things being particles of a real and rhythmic whole. We shall be able to communicate with one another instantly, irrespective of distance. Not only this, but through television and telephony we shall see and hear one another as perfectly as though we were face to face, despite intervening distances of thousands of miles; and the instruments through which we shall be able to do this will be amazingly simple compared with our present telephone. A man will be able to carry one in his vest pocket.*
>
> Nikola Tesla, *Collier's Magazine*, 30 January 1926

From Unwired to Wired to Wireless

The importance of technological change is not about the ability of humans to create some new device or process, but how an innovation meets changing social needs and then serves to transform that society. At the end of the 19th century, the world had an increasingly interdependent economy, with commercial, financial, military, political and colonial competition and cooperation across great distances. The needs and rewards for faster communication were substantial. The century had seen the electric telegraph reduce the distance between people; the telephone had begun to affect some aspects of life. But when the potential of radio waves was unlocked, decade by decade, it came to transform human lives in numerous ways which we now take for granted. In tracking the ideas, the technology and the applications of the era of wireless communication using radio waves, we can begin to see from our experience in modern history how different, complementary and cumulative can be the impacts from a single origin.

Until the 19th century, communication between individuals at home or work, military commanders at war and administrators across empires were limited by the speed of land transport or sail to carry written correspondence, or by lines of sight to transmit messages. Then, for a few decades the electrical telegraph, followed by the voice telephone, provided means to supplement this, before the beginnings of the wireless revolution. Reflecting a rapid succession of experimental stages, it was clear that wireless technology would form part of the future landscape when first patented in 1896, and the first demonstration of wirelessly transmitted speech and music was just ten years later in 1906. Just a century separates the practical origins of radio transmission in the 1890s and the first smartphone in the 1990s.

That period had seen the rapid extensions of experimentation into widespread applications. Maritime needs were at the forefront of wireless use: providing links from ship to shore, aiding in safety, rescue, weather information and navigation. Radio communication made its mark on land and in the air in the First World War. Colonial administrations were now in contact with imperial governments, and international news services had new facilities at their disposal. From the 1920s radio ('the wireless') could bring entertainment and information into the home, indirectly serving to reinforce regional and national identities. Television transmissions began before and resumed after a Second World War in which radar was a crucial weapon. Those who constructed their own crystal set as teenagers in 1928 to catch early radio broadcasts would own a pioneering mobile (cell) phone in their senior years and might live long enough to use a smartphone to see and speak to family on the other side of the world or view a distant sports event. The wireless revolution would transform almost every aspect of human interaction and society, from finance and business to political propaganda and the control of crime. Communication had ceased to be a matter of space. Wireless communication was a revolution with as important transformative impacts as any in history.

The current speed of technological change which affects work, study, recreation and social life can take our attention away from how dramatic were the beginnings of wireless technology and its ability to transform human society. But key transformations like those in the wireless revolution and the introduction of the printing press can span one long lifetime. The sequence of stages of the wireless revolution mapped out in this chapter will probably be seen just as a single episode by future historians, a brief transitional stage. The long history of development of communication is far more complex.

Foot Transport and Horse Messengers

The first human communication was of course by personal interchange, with the rate of transmission of information controlled by the pace of the human step. A normal walking pace is something like 5 km (3 miles) per hour, with a journey of 30 km (under 20 miles) in a day for those walking over even ground. Information would pass between communities in their periodic interactions, and the trade of knowledge and news like the trade in goods could move slowly across wide areas.

The despatch of urgent or important messages required speedy and dedicated messengers. As mentioned in Chapter 3, in 490 BC the herald Pheidippides ran with the urgent message to tell the citizens of Athens of their success in the battle of Marathon, whose name we have transferred to a sporting race of this length; the story of his death upon completing the journey of about 40 km (25 miles) indicates the stress that could come with this role.

The key to rapid communication lay in relays of couriers who could convey their message limited only by the running speed of each individual courier. Unlike the great civilisations and conquerors of the Old World, the Inca in and beyond Peru lacked an animal on which a messenger could ride. Inca imperial administration between the 13th and 16th centuries AD required liaison across a huge area, and the vast 40,000-kilometre network of roads and paths in Inca territories allowed relays of runners (*chasquis*) to convey messages long distance to maintain and reinforce central control over a broad area. These fit messengers could pass on oral information, carry the *quipu* of knotted string which encoded those messages and interpret the *quipu* when required. Spanish records claimed a journey of 200 km (125 miles) or even more could be covered in 24 hours by a sequence of fast runners travelling through the night as well as the day. Such a system, of course, required housing and provisioning for the network of couriers.

But foot couriers were not needed in the Old World once couriers were mounted on horses (or camels), with messages and news not limited by the human pace. While a walking pace of a horse is about 7 km (4 miles) per hour, a consistent trot could be 13 km (8 miles) per hour. A sequence of couriers riding at a fast pace with changes of horses could travel perhaps 160 km (100 miles) or more in a day, especially if the journey continued on a moonlit night.

The 2,700 km (1,700 miles)-long Royal Road in the reign of Achaemenid King Darius (522–486 BC), with rest stations at intervals of approximately 17–24 km (11–15 miles), supported relays of *pirradazish* horseback couriers.

A message from the capital Susa in Persia as far as Sardis in western Anatolia could have taken just twelve days.[1] This enabled the Achaemenids to rule a vast area and to signal the need to move military resources as required.

Rome's empire required communication between the imperial capital and military leaders in the field. Normal messages could travel along the network of roads by a relay of horseback messengers, or an important courier might be conveyed in a carriage with a regular change of horses. A message by the regular mail service, the *cursus publicus*, might travel by road 60 to 75 km a day; an academic calculation is based on a figure of 56 km (35 miles) per day.[2] One estimate suggests thirty-nine days' journey by land from Rome to Cordoba in Spain. But the Roman imperial territories lay in an arc around the Mediterranean Sea, and it was under sail rather than by road that the fastest movements of information and goods took place.

The Mongol Empire in the reign of Chinggis (Genghis) Khan (1206–1227) used horseback messengers to manage a system of administration and communication. The main routes used by the Mongols had regular stations (one estimate is 10,000) where a rider could exchange his horse for another. Marco Polo was especially impressed by the system and the speed by which it could move messages across the Mongol-held territories. Routes which initially linked Mongolia with Central Asia would in time extend eastwards into China.[3]

The Pony Express of the United States conveyed mail across the continent in 1860–1861, until soon eclipsed by the telegraph (Figure 31). The journey followed a route estimated as just less than 3,000 km (1900 miles), with stations for a rider to change horses every 16–20 km (10–12.5 miles). Using fast light riders and limits to postal loads, it was possible for a sequence of two riders to achieve 230 km (140 miles) per day (riding through the night) and cross the continent in little more than ten days.[4]

The stage coach or carriage was a more common means of conveying mail in Europe before the motorised era and would suffice for non-urgent messages. The speed was determined by the weight of the wagon and the passengers and goods being hauled, as well as the condition of the roads being used. Dedicated mail coaches, such as used in Britain from the later 18th century, helped guarantee a speedier delivery. The coach or carriage was more cost-effective than a rider on a horse, as personal and commercial communications could be carried for an acceptable fee.

Where horses were not available or suitable, messages could of course be sent by people using other domesticated animals. Chapter 3 mentioned haulage of carts by oxen, mule, and donkey. And notably the camel served

FIGURE 31 Pony Express stamp, 1861 (public domain).

ahead of these especially in arid areas, as used in the Middle East, Central Asia and the Indian subcontinent. A camel courier could if required cover up to 145 km (90 miles) in a twenty-four-hour period.[5]

Ships and Colonial Empires

Communications on land – whether on foot or by horse or carriage – depended on the range and quality of roads and tracks. A land-based empire such as that established by the Incas or Persians or Mongols, as with the Romans in the interior of Europe, might construct such a network for administration, but initial military campaigns of conquest depended for their communication on tracks created by those they conquered, or on travel across open land. Roman roads helped the empire retain its conquests against rebellion – not achieve them.

River vessels had long supplemented and bettered the speed of land transport and communications. The great civilisations of the Nile, the Tigris and Euphrates, the Indus, China's Huanghe (Yellow River) and elsewhere used these natural arteries. Access to the seas had extended this power. Maritime communication required no management of routes, only of ports. The speed of communication relied on the quality of the vessels and, in the age of sail, on the winds and currents. Much of Rome's empire was in the lands around the Mediterranean, and the speed of vessels sailing on the Mediterranean played at least as important a part as roads in centralising imperial administration and military control across the empire.

Roman ships, powered by sail and oarsmen, could travel at up to 4 to 5 knots, equivalent to 175 km (110 miles) and more over twenty-four hours of sailing.[6] One estimate assumes an average summer sailing speed across the Mediterranean of 3.05 knots (5.6 km or 3.5 miles per hour). A journey from Rome's domestic port Ostia to Gades in southern Spain could have taken 14.5 days sailing, 140 km (85 miles) a day; Ostia to Alexandria 20.5 days at a slightly slower pace. The maritime journey to Constantinople, linking the centres of the western and eastern Roman empires, would be of similar time by sea, compared to seventy-four days if done by road alone. Even so, taking land connections into account, messages sent between Rome and Constantinople could require some fifty days for a reply.

With the exception of Russia, whose territorial acquisitions controlled contiguous areas of the Eurasian landmass, the European empires of the modern era were widely spread and required maritime communication between metropolis and colony. Sailing vessels conveyed information and messages as much as troops, civilians and goods between the trading stations and overseas possessions of the Portuguese and Spanish, Germans and Dutch, Danish, British and French. Reporting to European capitals and receiving advice and instruction from there would be an essential part of European colonialism. The Archivo General de Indias in Seville contains 9 kilometres (5.5 miles) of shelves for records relating to the Spanish overseas possessions. Britain's India Office records span 14 km (9 miles) of shelving.

Britain's East India Company operated trading stations on the coast of the Indian subcontinent from the early 1600s and after the Battle of Plassey in 1757 began to acquire control of territory. Until the 1830s, communications between London and India were carried by vessels sailing round the Cape of South Africa (held by the Dutch from 1652 and, from 1806, the British). This was a maritime journey of four to six months, dependent on winds and the time of year. An exchange of messages could take a year or even more. The option of a land route through the Middle East was tried for communications between Britain and India but was unreliable.

The relationship between colonies and imperial capitals changed with the development of steam-powered sea-going ships, which can be traced back to 1813 with tests and trials which resulted in commercial crossings of the Atlantic by the late 1830s.

From 1835, steamships were sailing between Bombay and Suez, and mail transited overland through Egypt to the Mediterranean for onward transmission to London, reducing the one-way despatch times to a possible two months.[7] By the 1850s, the regular P&O ships meant the one-way transmission of mail could take six weeks or less – plus onward local time within India – thus allowing an improvement on the efficiency of exchange. The Suez Canal, opened in 1869, reduced the journey so that a 1913 trip from London to Bombay could be achieved in thirteen days by train and steamship.

Colonial administration had changed from an era of independent judgements by those with military or civil responsibility to an environment in which the government in Europe could expect to be consulted and to advise on decisions in a relatively timely fashion.

Semaphore

Distance signalling by different means – what we call semaphore – overcomes the limits of physical movement by humans by water or on foot or with domesticated animals. As mentioned in Chapter 2, fire can be used to send a message. Signals by beacons supplemented the messengers on the Persians' Royal Road. A fire signal announced the fall of Troy in Aeschylus' drama *Agamemnon*. A sequence of bonfires (or fire beacons) was set up on the English coast to warn of the arrival of the Spanish Armada in 1588.

Unlike fire beacons, smoke signals were useful only in daylight hours. Towers along the Great Wall of China could send messages ahead of those conveyed by human hand, with smoke colour-coded to indicate the level of a threat. The image of Native North Americans using smoke signals is one that resonates with us today. For all these methods, a limited number of messages are possible.

Far more varied communications could be sent with the use of visible 'arms' or flags or flashes of light linked to a linguistic code. During the revolutionary period of France, Claude Chappe experimented with methods of sending semaphore messages with his optical semaphore telegraph between mutually visible towers at 10-km (6-mile) intervals. The variable positions of arms conveyed the text of a message with ninety-four possible symbols. When a symbol represented a full code word, a long message could be transmitted. Such a system required good weather for visibility and could send two symbols a minute. The French semaphore network extended out of Paris, beginning in 1794. There was some caution about its security. Napoleon authorised the weekly transmission of winning ticket numbers in the national lottery.[8] In 1837, semaphore telegraphy was made a government monopoly, reaching many French cities by 1846 with almost 5,000 km (3,000 miles) of communication lines.

Other European countries developed more modest versions (Figure 32).[9] Abraham Edelcrantz in Sweden was working at the same time as Chappe, using shutters rather than arms with the advantage that these could be illuminated to allow night communication. Begun in the 1790s, this system was still operating in parts of Sweden in 1881.

Much naval communication was assisted by the use of flags – the simple code of raising them on the rigging of a ship, or holding semaphore flags in pairs at angles to represent individual letters or numbers. Ships could send messages at night by a heliograph such as the Aldis lamp – flashing lights

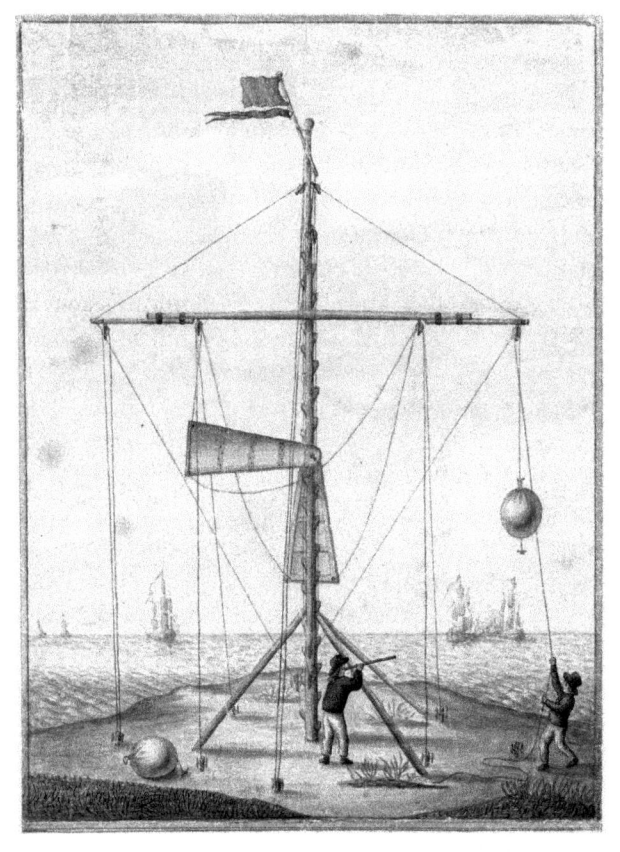

FIGURE 32 Semaphore signalling at Scheveningen, 1799 (Rijksmuseum, CC0, via Wikimedia Commons).

which used code for letters. Even after the introduction of wireless communication, night-signal lamps still found their uses.

Pigeons

A fascinating sideline in communications history is the messenger homing pigeon. While highly focused technological innovations were constantly improving both the means of modern military communication and the means to monitor the wireless messages of the enemy, over the heads of the armed forces of the 20th century flew small birds carrying written (if necessarily coded) messages.[10]

The speed (typically 60 km or 35 miles per hour) and range of a trained messenger pigeon is remarkable.[11] A normal flight up to 32 km (20 miles) could be extended at extremes to 160 km (100 miles) or even more, including flying at night by pigeons trained to do so. The core principle is that a pigeon is taken away from its home (or a new location it is trained to consider as home) so that when released by the sender it returns there, carrying a paper message as required. Among pigeons' advantages in wartime over human messengers, 'they cannot share secrets or act as double agents'.[12]

There are early references to using birds for messages in the ancient world. Pigeon messengers were used in the Roman Empire: Roman author Pliny noted a message to the consuls was sent by pigeon in the battle with Mark Antony in 43 BC. Crusaders in Palestine used falcons to try and catch pigeons used by the Muslim armies to communicate.

In modern warfare, pigeons' use has been as an alternative communication when radio was unavailable or insecure. Their value was shown in the 1870 siege by the Prussian forces of Paris, which inspired pigeons' use in Europe to send news and even stock-price information. In the First World War, French, British, German and then, from 1917, United States forces used pigeons; despite advances in radio technology, tanks were unable to communicate by radio, and pigeons could serve instead. An estimated 500,000 birds operated during the war years.[13]

Uses in the Second World War were more specialised, but the United States used 54,000 pigeons and the British 200,000 pigeons. The first news of the success of the D-Day landings was carried back to Britain by an Irish pigeon called Paddy. The US military continued to use pigeons in Korean War areas where there were territorial barriers to radio transmission, until the US Army Signal Corps Pigeon Service finally ended in 1957. Even by 2012, the government of China noted it had 50,000 trained pigeons to supplement its high-end military technology.[14]

Telegraphy

The term 'wireless' is significant because it marked the break from *wired* communication – the telegraph cable (which itself had only a history of some six decades before wireless alternatives) and the wired telephone, whose commercial use anticipated radio by only a couple of decades.

As with most scientific advances, the commercial and administrative uses of the electric telegraph followed a long sequence of innovative and creative research and invention. At least sixty experimental attempts in telegraphy had been undertaken before the mid-1830s.[15] The development of the railways stimulated a commercial application. Britain's Stockton and Darlington Railway had opened in 1825; because trains travelled in both directions over a single stretch of rail track, real-time information was required along the length of the lines to check if the route was open and safe to use. Telegraphy lines were installed alongside railways from the 1830s, and it was these which could initially provide communication for wider use. Among competing innovators, in 1838 Briton Edward Davy (1806–1885) patented a system of wired communication which was valued for the practical application of messaging on telegraphy cables extended along the railway lines. Public awareness of the value of this system was demonstrated when the early telegraph on the line between London Paddington and Slough stations was used to alert police about criminals who had boarded the train. Pickpocket Fiddler Dick was arrested by this means in Slough in 1844. In the following year, Slough police could advise their London counterparts that murderer John Tawell was on the train to Paddington, leading to his arrest, trial and execution. By 1848, half the railways in Britain had a telegraph wire alongside the track.

The telegraph moved messages quickly between towns, but they still had to reach their final addressee. Speed of final delivery was particularly important in financial affairs, and this was achieved by the development of pneumatic tubes in which air and vacuum could be used to transmit a message from telegraph office to a stock exchange, as implemented in London in 1853. Pneumatic tubes continued to be used well into the later 20th century for internal communications, with mechanisms to identify the destination station.

An independent telegraph line was established in the United States when federal funds were granted to Samuel Morse (1791–1872) to create a connection between Washington and Baltimore, operational in 1844. Different lines were developed in and between eastern cities from 1845, reaching west to cross the Mississippi at St Louis by 1848, and completing the transcontinental line (to replace the short-lived Pony Express) by 1861. Four years later in 1865 (following the failure of an earlier cable which broke), secure telegraphy was achieved by transatlantic cable between North America and Europe. Individual countries developed their own telegraph systems, with France and England linked in 1850 (Figure 33). Britain made telegraphy a state-run monopoly in 1870.

After different approaches to communicating words by electrical pulses had been explored, the most effective proved to be that developed by Alfred Vail (1807–1859) in 1840 and associated with the name of Morse. In this, an operator would be trained to code letters with a combination of short and long signals ('dots and dashes') on the telegraph key. The combinations chosen were related to the frequency of letters used in typesetting: the most common letters in English having the shorter, simpler codes.[16] By the mid 1850s, the message could be printed out without the need for a second operator to recognise the code and record the words.

The value of rapid messaging in military campaigns became increasingly clear; the Prussian military had access to a system from 1847. As the Crimean War of 1853–1856 unfolded, different participants set up field

FIGURE 33 Goliath steamer laying telegraph cables, 1850 (photo credit: 19th era 2/Alamy Stock Photo).

lines for communication: the Russians in 1854–1855, followed soon by the French and British.[17] An adapted plough was trialled, with only modest success, to run a cable and bury it underground. As would be found elsewhere, military commanders making decisions in the field were less than enthusiastic about the ease with which messages, advice and instructions could be sent to them from London or Paris.[18]

The electric telegraph was a primary weapon in the 19th-century conflicts in North America.[19] Its importance loomed large in the years of the Civil War (1861–1865), to the benefit primarily of the Union as the United States Army Signal Corps was formalised in 1863. Some 40,000 km (25,000 miles) of cable lines were laid by the Union side, 13,000 km (8000 miles) by the Confederates. An estimated 300 telegraph operators (often young boys) died in the conflict. As the power of the United States spread west into lands of Native Americans (with telegraphy cables, as in Europe, often following the lines of railways), military and civil communication by telegraphy gave the United States forces advantages their Indigenous resisters could never hope to match – although by sabotaging the line, these could delay their opponents.

The electric telegraph had a major impact on colonial administrations; it was used by Britain in the Ashanti War in West Africa in 1873 and the Anglo-Zulu war of 1879, and the Sudan was linked to Cairo by 1870.[20] By revolutionising the means by which colonial military and civilian officials communicated with each other and with the central colonial governments, the telegraph transformed their relative power. No longer need (or could) the man in the field take on himself the full responsibility for a military or administrative decision: London or Paris or Brussels or Berlin could be consulted and could send instructions. This placed the authority for a decision back in the relevant colonial office of the central government; something those in the field may at times have welcomed but may also have regretted as their advice was ignored by distant senior officials.

Internal communication in India had been strengthened after the 1854 Post Office Act by mail boats and dedicated mail carts to supplement the earlier pattern of runners carrying mail.[21] Railways had also begun in use for regular mail from the 1850s. The same British administration, under the Marquis of Dalhousie, had in 1850 sponsored an experimental telegraph line and in 1853 began constructing the new telegraph between major Indian cities, demonstrating the value provided by the ability to send messages and receive replies in a day instead of a month, as previously.[22]

The Indian Mutiny of 1857 provided a demonstration of the power of the telegraph.[23] By this means, British military commanders could be in contact with each other and with the central civilian administration (then under the control of Britain's East India Company) after the outbreak of the rebellion by Indian troops. Without this tool, we can assume the conflict would have continued longer, even if with the same eventual result of crushing those who rebelled against British power. 'The Electric Telegraph has saved India' was the claim of colonial official Sir Robert Montgomery – an observation so significant that it was inscribed on the Telegraph Memorial erected in Delhi in 1902 'to commemorate the loyal and devoted services of Delhi telegraph office staff, on the eventful 11th May 1857' when scattered British military stations were informed of the outbreak of the mutiny.[24]

Noting the success of the telegraph in and after the war, India's administration, now under the direct control of the British government, expanded the lines for the telegraph system within India – 28,000 km (17,400 miles) by 1865, 84,000 km (52,000 miles) by 1900, linking almost 5,000 telegraph offices.

In 1865, a land telegraph line from Istanbul to the Persian Gulf provided a link between Europe's telegraph lines and India, although with multiple relay stages which meant a message could take more than six days. The 1870 marine cable between Suez and Bombay reduced to six hours the time to receive a message from London and provided the security of being all under British control, although there was a significant cost for each telegram sent. The experience of the telegraph in India inspired colonial powers to improve their links both to colonies and to overseas dominions. By 1872, Australia was first linked to Europe by underwater cable from Dutch Java.

One important benefit of the telegraphy system was in the transmission of news. News agencies developed to supplement individual foreign correspondents, beginning with the German firm Reuters from 1850, which opened its news wire service from London in 1851. Reuters had earlier used carrier pigeons to send business information to customers. Both news reports and military messages had used telegraphy in the Crimean War of 1853–1856, demonstrating its value to those participating in or guiding world events, and to those who wished to be aware of them. This would bring early news both good and bad: news of successes and of defeats. It had taken six weeks for London's government to learn of the Declaration of Independence in their American colonies, and twelve days to learn of the

British victory at Trafalgar. Now information could travel to allow, and require, immediate response.

The Wired Telephone

The wires of the telegraph transmitted messages by electrical impulses which required human or automatic action to translate them back into text. The challenge to transform speech into electrical impulses which could be heard immediately as speech by the listener had also involved many competing attempts.[25] In 1876 – some thirty-two years after Samuel Morse's telegraph message to Baltimore – Alexander Graham Bell (1847–1922) was able to demonstrate and patent a successful result of experiments and developments. The more widely used term 'telephone' then became applied solely to this system. Many contributors were involved in the subsequent development and refinement of the telephone system.

Telephone links were initially constructed to serve and link private customers within a limit of a little over 30 km (20 miles), a Boston businessman connecting home and factory in 1877 being a pioneer.[26] Telephone switchboards linking a small number of subscribers followed. By 1880, there were 30,000 telephones in use worldwide. Only in the following decade would wider networks be developed, beginning in the eastern United States. Between 1895 and 1905, the number of telephones in the USA grew from 340,000 to over 4 million.[27] Even in 1933, 53 per cent of world telephones were in the United States.[28]

Sweden and Norway developed telephone systems sooner than other nations of Europe. Britain's telephone system began in 1878–1879. There was initial caution, with engineer William Preece of the Post Office suggesting in 1879 'I fancy the descriptions we get of its use in America are a little exaggerated; but there are conditions in America which necessitate the use of instruments of this kind more than here. Here we have a superabundance of messengers, errand boys, and things of that kind.'[29] Private companies did develop, operating under licence from the Post Office, and telephone links to France were established in 1891 by submarine cable, just a year later than links between London and the British Midlands.[30] In 1912, telephones became a unified state-owned operation. The telephone did not make telegrams redundant: in 1918, some 82 million telegrams were sent in Britain.

Only in 1927 was a commercial telephone link between Britain and the United States established, long after the revolution of wireless communication had begun. Australia's direct telephone link to Britain had to wait until 1930, by when radio communication was already well established.

Communication without Wires

In the long extent of human history, the most dramatic transition in communication was from a world where this could only be between humans near (or at least in sight lines of) each other to a world where radio waves were harnessed to transmit messages, news, entertainment and more across vast distances. These two stages were separated by only a few decades in which human communication had been advanced by the use of wired systems for electric telegraphy and voice telephone. Telegraphy began to spread in the 1840s, telephones from the 1880s and wireless communication began in the decade of the 1900s as ships were linked by radio to land. By the 1920s, domestic radio was widely available to the public (and the first stages of television developed in the 1930s).

As with wired telegraphy and telephone, many individuals were involved in experimental work that led to the rapid development of forms of wireless communication.[31] Once the potential power of harnessing radio waves became clear, multiple applications developed rapidly with further research and both commercial and government investment.

Key among the advances in theoretical physics that underlay these developments, the Scotsman James Clerk Maxwell (1831–1879) played a major role, marking his work out as so revolutionary it has been compared with that of Newton and Einstein in its effective impact. Maxwell's studies included a major paper in 1865 'A Dynamical Theory of the Electromagnetic Field', and in 1873 his book *Treatise on Electricity and Magnetism*.[32] His presentations brought together into a unified model concepts and understanding of the relationships between magnetism and electricity and therefore the nature of light. This included analysing the nature of electromagnetic waves and demonstrating that these all travel at the speed of light, varying in wavelength and frequency. The theoretical basis was thus set for further study of radio waves.

The German scientist Heinrich Rudolf Hertz (1857–1894) supported Maxwell's theory in his experimental work, notably in the years from 1886 to 1888 with the generation, transmission and reception of electromagnetic waves in the radio-wave range of 50 to 500 MHz.[33] He died young, before he was able to see that he was mistaken in his doubts as to whether radio waves could be used in wireless telegraphy.

While Hertz was not focused on practical uses, others addressed the potential application of the ability to generate and transmit radio waves: those electromagnetic waves whose wavelengths within the electromagnetic spectrum are longer than infrared light and with lower frequency.

There were many contributors and many researchers in experiments with radio waves and wireless transmission from 1891 onwards, working simultaneously in different countries.[34] German Erich Rathenau sent a message over a 5 km (3 mile) distance in 1894 and Karl Strecker over 17 km (10.5 miles), both by conduction over low ground-level surfaces.

While others would continue to make major contributions to technology and applications, notable in his lasting practical impact was the Italian Guglielmo Marconi (1874–1937) (Figure 34). His experiments with transmitting radio waves led to his patenting in Britain in 1896 'Improvements in Transmitting Electrical impulses and Signals, and in Apparatus therefor'. In the following year, he founded his Wireless Telegraph and Signal Company. He was able to demonstrate the effectiveness of his approach with a message sent over 29 km (18 miles) across the Bristol Channel.

As so often (we can think of Leonardo da Vinci), it was because of the potential value to the military that the first encouragement and developments took place. Henry Jackson of the British Admiralty had sent Morse code messages wirelessly a short distance by 1896.[35] Marconi's ability to transmit radio waves in controlled fashion had major advantages for naval vessels, for which the progress of earlier decades in wired telegraphy and telephony were irrelevant. After further experiments in 1898, the British Navy proceeded enthusiastically to encourage and embrace the potential of wireless communication.

Similar interest engaged the Russian Navy, which built on the radio-waves research of Alexander Popov (1859–1906).[36] Initially, Popov

FIGURE 34 Marconi in 1896 with his experimental radio equipment (Smithsonian Institution, public domain, via Wikimedia Commons).

constructed a transmitter, then a receiver, and in 1896 he transmitted radio waves an experimental short distance on land. This was extended to communication between naval ships in the following years, and by 1899 a message could be transmitted for 50 km (or 30 miles).

Serbian-born Nikola Tesla (1856–1943) worked in the United States more as an inventor and engineer than as a formal scientist. Working initially on developments in electricity, he undertook further investigations inspired by Hertz's work on radio waves, demonstrating their transmission through earth and space and testing transmitters. In 1898, Tesla showed how a model boat could be controlled by radio waves, and he filed a patent for this process that year. In 1899, he was experimenting with wireless telegraphy from the heights of Colorado Springs. He filed a US patent on a 'System of Transmission of Electrical Energy' in 1897, granted in 1900, and this inhibited Marconi from establishing patents when he applied later that year (although in 1904 the Patent Office decided in Marconi's favour). Tesla continued in competition with Marconi, but the systems and businesses of the latter were to dominate radio development.[37]

Extensions of experimental communication now developed rapidly with maritime uses at the forefront. Morse code found a new application, and the distances achieved by transmitters grew. In 1901, Marconi was able to send a transatlantic message 2,700 km (1,700 miles) between Cornwall in England and Newfoundland in Canada. With rapid take-up in maritime contexts, the International Radiotelegraph Convention in Berlin in 1906, at which twenty-seven countries were represented, agreed that the Morse code for SOS (three short, three long, three short) would be the standard emergency call signal. Government control of the new technology was inevitable; Britain's Wireless Telegraphy Act applied from 1904.

Marconi's imaginative and creative approach to the application of radio waves was recognised by a Nobel Prize for Physics in 1909, but he shared that with the German academic scientist Karl Ferdinand Braun (1850–1919), who had worked on transmitters and receivers of radio waves.

Morse code had its value, but given the success and appeal of the telephone, research developed into wireless speech transmission. The Canadian/American Reginald Fessenden (1866–1932) had worked on the transmission of Morse code pulses by wireless telegraphy and made major contributions in this area; but he also explored the potential of voice transmission (Figure 35). In 1900, a (distorted) voice transmission provided reassurance and encouragement, and in 1901 he submitted a US patent application based on his work. By 1904, Fessenden transmitted sound using radio waves over a 40-km distance. A demonstration in Massachusetts in

FIGURE 35 Reginald Fessenden working on radio voice transmission (photo credit: Pictorial Press Ltd/Alamy Stock Photo).

1906, transmitting speech and music across a distance of 18 km (11 miles), was of more encouraging quality (sometimes called the world's first radio broadcast), and he achieved significantly greater distances in the following years: 1,000 km (600 miles) by 1908.[38]

The future potential of voice conveyed by radio waves was established. By amplitude modulation (AM), sound was converted into electromagnetic waves, which were reinterpreted as sound by the receiver. Used in the first entertainment radio systems from the 1920s, FM (frequency modulation) resulted from research in the 1930s and would follow much later in the use of entertainment radio.

The challenges and demands on the nations in conflict during the First World War stimulated not only further reliance on wireless communication at sea, but also initial developments in other contexts. As this chapter goes on to discuss, battlefield dependence on wired telephone and telegraph cables was supplemented by radio. Those in aircraft and airships saw the first uses of radio both for communication and for determining their position. Control and sabotage of submarine cables led to the use of wireless communication for messaging, news and propaganda.

While telegraphy had the advantage over cables that it could not be cut by enemy action, the radio relay stations certainly could be attacked, and both sides in the war used their navies to attack shore-based wireless stations.[39]

After the end of the war in 1918, attention turned to the wider social potential of wireless communication and the ability to convey sound (whether spoken word or music) to a broad audience who possessed their own receivers at home. The terms 'radio', 'wireless' and 'broadcast', like the later term 'television', were now commonly applied to the domestic setting. The entrepreneurial commerce of the United States led the initiatives. There had been amateur enthusiasts experimenting with their own private networks; before the war an estimated 5,000 radio licences had been issued by the British licensing authority, the Post Office. Some American universities had internal broadcasts of news and weather before the First World War.[40]

The Westinghouse Electric and Manufacturing Company established Radio KDKA, which began broadcasting in Pittsburgh in 1920. By 1922, radio in the United States was a $60 million business.[41] Radio would be a competitive commercial landscape in the US.

By contrast, radio in Britain was a government-controlled and funded exercise. The first radio broadcasts by the newly formed British Broadcasting Company (BBC) were in November 1922. News was an important part of the role of the BBC, exemplified in the crisis of Britain's 1926 General Strike. The rapid spread of radio was stimulated by the low cost of entry for those who could make their own basic receivers. The 'crystal set' was readily constructed by young and old. My own father, born in 1914 before the birth of broadcasting, constructed his crystal set as a teenager; then, just seventeen years after the beginnings of radio, presented his own first talk on the BBC Midland service in July 1939.

The rapid speed of development in wireless communication saw experiments with television begin only shortly after public voice radio, but the economic downturn of the 1930s followed by the outbreak of the Second World War delayed the spread of commercial television. John Logie Baird's experiments with television from 1925 showed what could be done. In 1928 in the US, television pictures with sound were broadcast, and there was a gradual but small sale of television receivers by 1939. Britain's BBC experimented with television broadcasts for subscribers between 1937 and 1939. The post-war revival of television was rapid, with a million sets in use in the United States by 1948; by 1952, there were somewhat fewer than 3 million sets in Britain.[42]

Meanwhile, other potential uses of harnessing radio waves were being explored and developed. A survey of the many impacts of radio technology emphasises how dramatic has been the wireless revolution.

The Changed Wireless World

Once the physics of electromagnetic waves had been understood, the way was open to consider technological and commercial adaptations of this new knowledge. But inventions and adaptations do not come about automatically or inevitably. It was not immediately obvious to the earliest experimental scientists what would be the advantages to society from radio waves. These arose from the combination of entrepreneurial individuals such as Marconi, commercial imperatives in consumer entertainment radio and the advantages to specific groups, including the maritime community, the military (accelerated by the demands of the First World War), news media, financial institutions and government agencies. The multiple advantages of a wireless world would become clear, despite the initial view in some quarters that the achievements of the 19th century in cabled communication and telephony were sufficient.

As with other major innovations discussed in this book, in order to demonstrate the very many significant changes achieved by control and use of radio waves, we need to strip away the comfortable familiarity we have and reposition ourselves in the pre-wireless era. In 1900, civil, military and community links were set by the technology of the fixed cable, and access was limited to those in the times, places and economic position to benefit from these. Physical proximity set limits on entertainment. Within half a lifetime, access to the rewards of a wireless world achieved a transition that we see today as the norm. As described in the following account, the list of early rewards of the wireless revolution was dramatic.

Naval Communication

The importance of radio communication to mariners was revolutionary, and the stimulus from the maritime sector was crucial in the early development of wireless. No matter how efficient and widespread was the development of the cable system in the 19th century for despatching telegraph messages, and the telephone system to transmit voice, this was of no value to those at sea – the naval marine, the merchant navy, ocean liners or smaller vessels sailing the ocean to fish or ferry passengers or as offshore pleasure craft. A major incentive to fast track the development of wireless communication came from the needs to the maritime sector and for those on ships to communicate with each other and with the land.[43] Amid the early debates on the value of wireless communication, the advantages provided to those at sea loomed largest – and applied across a range of maritime communities. So it is natural that an early account of radio technology, A. P. Morgan's 1912 book *Wireless Telegraphy and Telephony*, should feature a link between ship and shore on its cover.[44]

Naval power was of great importance in the later 19th century, not least to the colonial powers.[45] An island nation like Britain might rely on its navy and their officers more than, say, Russia with its continental-based empire. The term 'gunboat diplomacy' seems to date from 1835. But naval power related not just to competing states with far-flung empires and dominions; it would prove relevant closer to home in the tensions between European states (including the Ottoman rulers in Istanbul) which led to the outbreak of the First World War in 1914. Those battleship commanders who could readily communicate with each other in the context of battle, and those governments and naval authorities who could send orders and information to their ships at sea, would acknowledge the major changes and strengths supplied by the wireless revolution.

The challenge to any navy was how those in charge of vessels at sea could keep in contact with each other and, equally important, with land-based authorities. Ships within sight lines could communicate by signal flags, or with semaphore using a heliograph to send Morse code messages, but when a fleet was in active service or away from its home base on a tour of duty at sea, no level of sophisticated progress in wired land telegraphy was of any use. So while there may have been government, army and commercial leaders who saw little immediate advantage in radio technology, those involved with navies and with the development and use of naval strategies and maritime forces could see a world-changing advantage in being able to communicate with warships wherever these might be.

Experiments by the British Navy in 1898 proved very positive, with radio equipment installed on three warships the following year and then a rapid expansion of ship-to-shore wireless communication.[46] At the end of 1900, the British Navy had 51 radio sets in use on ships or on shore; by the beginning of 1914, it had 30 shore stations and 435 ships fitted with radio apparatus. A dedicated naval unit, the Wireless Telegraphy Branch, was established in 1906 by when ranges for transmission up to 2,000 km (1,250 miles) had been achieved. This was well ahead of most

European competitors, and a contrast with the slow interest shown by the Army.[47]

Russian warships, building on the work of Alexander Popov, were equipped for wireless communication by the time of the 1904–1905 conflict with Japan, whose navy had also acquired radio technology. The potential was taken up elsewhere too, with very basic radio transmitters fitted in the American ship SS *Paul* in 1899 and the German SS *Kaiser* the following year.[48] In 1905, the United States Navy succeeded in sending a radio message 3,500 km (2,150 miles) between stations at Coney Island and Colon in Panama.

Warships linked together and to land command by radio played a major role in the First World War. This included patrolling sea lanes to inhibit supplies to the enemy, in which the Allied action against Germany was largely successful, blocking North Sea access to German ports. German submarine U-boats in turn attacked ships supplying Britain across the Atlantic. Land-based intelligence could advise ships at sea where to focus their attention, and ships could report the outcomes of their manoeuvres. The failed Gallipoli campaign of 1915–1916 by Allied forces against the Ottoman Empire sought to gain maritime access between the Mediterranean and the Black Sea where much of the Russian fleet was based.

The threat from U-boats was one stimulus for research into sonar technology – the ability to send underwater sound waves which would bounce off an underwater surface as a trackable and measurable echo. Reginald Fessenden patented a system in the United States in 1913, and locating an iceberg at a distance of 3.4 km (2 miles) in 1914 demonstrated the marine safety potential.[49] Developments of devices to track submarines during the First World War were completed only by 1918, by the United States, France and Britain. Radar – using radio waves rather than sound waves to locate and track a distant object – would take this goal a major stage further from the mid 1930s.

Radio-controlled weapons might be some way into the future, but the first experiments included that of 1904 when Briton Jack Kitchen suggested to the British Navy a radio control system he had developed for torpedoes.[50] The French inventor Gustav Gabet took this idea forward in 1909 with demonstrations of a radio-controlled torpedo in the Seine River. By the Second World War, this development was fully functional.[51]

Marine Safety and Weather Information

Whereas wireless communication was the focus of attention by navies because of its contribution to military power, the ability to communicate long distance while at sea had a special value to other vessels. For the first time in the millennia of maritime history, radio made it possible to raise an alert in an emergency in order to secure rescue by other ships.[52] The contribution of radio technology in saving lives gives it special esteem in 20th-century developments.

Ships equipped with the ability to despatch emergency messages needed, of course, another ship or land-based station also equipped with radio communication.

The first maritime radio distress call rescue took place in 1899, when the German steamer *Elbe* ran aground and the lightship positioned near Dover was able to report the accident to the coastal radio station on shore.[53] In 1909, the British *Republic* and the Italian *Florida* collided south-west of Nantucket, and a radio message could be sent from the *Republic* requesting assistance. This was picked up by a shore station in Massachusetts, who contacted other ships in the area, allowing all lives to be saved.

Perhaps the most famous incident was when the *Titanic* ocean liner on its first voyage across the Atlantic in 1912 hit an iceberg 600 km (375 miles) south of Newfoundland. The Marconi company had supplied its wireless telegraphy system together with two radio operators. Emergency radio messages were sent in Morse code after the collision and were picked up by radio operators on other ships, leading one to reach the site of the sinking and rescue some of the passengers. Ironically, the *Titanic* had been informed by radio from another vessel, the SS *Californian*, of pack ice in the area, but the message was fatally mishandled after receipt. The United States thereafter required all passenger ships to operate continuous radio communications. In 1914, two ships, the SS *Majestic* and the SS *Cedric*, were forbidden to set sail because an inspection showed their radio equipment to be faulty.

Coastal wireless stations around Britain were installed with support from major shipping companies and the insurance exchange Lloyds of London, and from 1909 coastal stations were taken over by the Post Office, although some later passed to naval control.[54] From 1919, all British registered merchant shipping needed a radio operator on board, and passenger ships were obliged to carry three radio officers. International agreement in 1929 required wireless equipment on any passenger ship or reasonably sized merchant vessel.[55]

Radio messages could announce an emergency and summon assistance, but wireless messages could also help avoid danger by advising threatening weather. The use of radio for sending weather forecasts and weather warnings was a much-appreciated marine innovation. Meteorological services on land had been established in the 19th century: that of Britain established from 1854 could advise mariners of impending gales by signals from coastal stations. In the United States, an army responsibility for weather forecasting was established by a federal law of 1870, and it became a civilian responsibility in 1890.[56] Reginald Fessenden in 1900 aided the US Weather Bureau in investigating the possible transition of their communications from cable telegraphy to wireless, and in 1904 the Navy took over these responsibilities.[57] Radio warnings from land to ships became routine, and from 1913 the US service broadcast daily bulletins whatever the weather. In 1905, the SS *New York* transmitted its own weather message, as did the SS *Carthago* the following year, warning of a hurricane off the Yucatan coast, and vessels could expect to receive information from different sources when storms approached. From 1911, the British Meteorological Office routinely issued weather warnings at sea by radio.

An additional but valuable early contribution of radio to the maritime world was the radio notification of an accurate time signal, replacing reliance on chronometers in the calculation of longitude for ships' navigators. After initial experiments from Washington, DC, in 1904 wireless time signals were sent from the Navy Yard in Boston. With a Canadian initiative in 1907, an automated daily time signal (informed by telegraph) was sent to ships at sea.[58] France followed with a time signal in 1910, sent from the Eiffel Tower, which was then the world's tallest structure. The once-a-day despatch echoed the long tradition of a time ball or noon gun informing ships in harbour. Time markers as a sequence of pips became a feature of the more widely accessible BBC radio from the 1920s.

Military Tool

Wireless technology would come to provide land based military forces with a significant new weapon to aid communication; early engagement with radio communication was seen in the Russo-Japanese war of 1904–1905 and in the Balkan War of 1912–1913. But it also had a significant disadvantage. While a wired message, like those sent by person (or indeed pigeon), could be directed just to the recipient (though at some risk of interception), messages sent by radio waves could more easily be picked up by others – by definition they were *broadcast*. There were thus major security questions relating to the military use of radio. At the very least, messages needed to be coded, and their coding and decipherment added time delays to radio messages.

The Changed Wireless World

Marconi himself was aware that military conflicts and military needs stimulate investment in technological innovation. The British Army (in contrast to the Navy) held a sceptical attitude to radio, with just a few unenthusiastic early experiments.[59] The mobility of equipment and limitations of transmission appeared practical barriers, with a conclusion that 'an inefficient wireless service is worse than none at all'.

When the South African War broke out between the British and the Boer republics in 1899, Marconi proposed to the British War Office that radio be used in areas where reliance on telegraph wires would be unrealistic.[60] Though his proposal was initially accepted, the experiment was soon considered unfeasible, and the equipment was dismantled by the army telegraph unit in February 1900.[61] The Marconi Company went on to develop the Crystal Receiver no. 16 in 1906, which cut down on static noise to improve reception.

To land forces in the First World War, telegraphy and telephone were key and formed an essential role on the front. But cables that might be used to link trenches and command centres could readily be cut by enemy action, and wireless alternatives began to play some part in land warfare.

As the advantages of wireless communication over telephony and cable telegraphy at the battle front became clear, Germany's military led the way in radio and entered the First World War with a commitment to wireless communication (Figure 36).[62] War priorities also led Germany to erect towers for wireless transmission within the territory of their wartime allies the Ottoman Empire.

At the outbreak of the First World War, the British authorities reacted with security concerns to render the radio equipment of local amateur enthusiasts inoperable. These individuals were subsequently brought back into the war to be ready to provide services if required. Volunteers from the British Post Office were called on for service at the Western Front.[63]

From 1915, the British began to take seriously the option of radio at the front. British radio field sets required heavy transport and could provide communication over only a few kilometres. The 'Knapsack Station' required a unit of four men to carry it and set it up. As the war developed, some improvements to the portability and range of radio were seen by 1917, and several key areas of fighting on the Western Front showed the British forces the advantage offered by wireless communication.[64] In more open fields of battle – the Middle East and East Africa – the installation of wireless stations had greater effect. Indeed, in East Africa wireless proved the main means of military communication.

FIGURE 36 A German field radio station in the First World War (photo credit: Chronicle/Alamy Stock Photo).

During the Russian Revolution of 1917 and subsequent civil war, the use of radio communication was crucial as the Bolsheviks sought to control a vast territory. News and propaganda bulletins were just as important as administrative and military directives, and a priority was placed on ensuring adequate facilities were constructed and maintained.[65]

With the beginnings of aircraft, the potential value for radio transmission from the air stimulated attempts to establish air-to-ground systems, though it was a slow process for an airborne crewman to despatch a Morse code message. Experiments with balloons in 1908 and a biplane in 1910 had shown the future potential of air-to-ground communication.[66] Pre-war German Zeppelin airships were then equipped with radio equipment.

In the First World War, with radio equipment weighing up to 45 kg (100 lb) and taking up the equivalent of an aircraft crewman's space, it had limited appeal.[67] In one role, when an aircraft flew over enemy batteries its crew could relay their position to ground artillery and then advise the level of success or otherwise of shells despatched. But gradually as the war progressed, the weight of radio equipment was reduced.

Late in the war years, Swedish-American Ernst Alexanderson developed a tuned radio-frequency receiver which could also transmit, powered by a dynamo driven by the aircraft propeller.

With the development of aircraft for non-military roles after the First World War, voice transmission between pilot and ground was a priority, and this was in use of planes travelling between London and Paris in 1919.[68] However, international agreement that passenger planes must carry wireless equipment was not in place until 1930.

Direction Finding and Navigation

Radio-wave technology was thus adapted early to serve maritime, aerial and land-based needs. Messages were passed between ships and land, between aircraft and land and within armies. Radio time signals could help ships confirm their longitude location. The development of directional aids to define location using radio technology provided major benefits for navigation on sea, in the air and even on land, in peacetime as much as wartime. Radio became a significant tool for mariners and aviators less than two decades after the first wireless transmissions.[69]

After partial success by others, Marconi patented a system in 1906 by which aerials could identify the direction of a signal, and the following year Telefunken in Germany developed a simple system in which an approximate compass bearing could be determined from comparing signal strengths from different transmitters.[70] As with other developments, many others (including Reginald Fessenden) experimented to improve the applications for direction guidance.

During the First World War, German Zeppelin airships carried bulky 'radio compasses' which received radio beams transmitted manually from Nauen in Germany (then the most powerful transmitting station in the world) and from Bruges in Belgium, allowing the airship crew to take a directional bearing and establish its own position.[71] The British found it possible to track such messages as they were sent and identify the imminent threat of enemy airships; they also found they could identify the location of enemy naval vessels by plotting their radio transmissions.[72]

Radio beacons transmitting at fixed times began to be used later in the First World War and provided aircraft with guidance on their position at night, especially important for night-time bombing raids.[73]

Radio as a navigational aid to shipping began early if selectively; in 1912, the British Cunard Line's large ocean vessel *Mauretania* had a direction finder.[74] From the 1920s, radio direction finding was actively developed and applied, following the same principle of using the comparative strengths of signals from shore transmitters (radio beacons) and triangulating from these to establish position.

Marine radio direction finding developed with three different methods in use.[75] Equipment on board a ship measured the direction and strength of shore signals to calculate their position. Alternately, shore-based stations could take bearings on transmissions received from ships and radio back to them their position. In a third experimental method from 1929, a ship could monitor radio transmissions sent from a shore station by a transmitter rotating at a known speed, which sent a distinct signal to the north, allowing the radio operator to use the time between the north signal and the weaker signals to give a bearing.

A more limited but important form of direction finding was used on major passenger routes in the 1930s. Radio direction antennae transmitted beams on major routes, and a pilot who veered off the expected route would be warned by a Morse code signal indicating whether he needed to adjust to port or starboard.[76]

The challenge for both marine and aircraft direction finding was the level of accuracy.[77] Mariners continued to use traditional methods long after radio was available, and there were continual changes and improvements in navigation equipment and directional transmitting beacons.

Derivatives of navigational system using transmissions from land-based beacons continued to be maintained until overtaken by satellite-based Global Positioning Systems, with United States military use from the 1970s and wider civilian use beginning from the 1980s, providing a new level of accuracy and universal access.

Radar

While ships and aircraft could use on board equipment to translate radio signals from onshore beacons to map their own position and assist them in navigation, this did not inform (or alert) those on land of their position. That role had to await the development of radar ('radio detection and ranging'). In this system, radio waves are reflected when they make contact with metallic objects, allowing the estimation of both the direction and distance of those objects (such as an enemy ship).

In the earliest years of research, scientists had noted the reaction when radio waves encountered solid objects, and experiments continued in Britain, Germany and the United States. A German patent by Christian Hülsmeyer in 1904 for identifying metal objects at a distance of up to

The Changed Wireless World

3,000 m (10,000 ft) was initially ignored by the naval authorities.[78] In 1917, Nikola Tesla outlined his ideas for the practical instruments needed for the development of radar.

The military value of such a system encouraged initiatives and developments of practical applications in several countries during the 1930s. British resources were put into developing technology which could detect aircraft and track their movement, with 1935 often cited as a turning point in this work. This would prove a major part of the British and Allied armoury in the Second World War. Radar was in use by the British Navy in the HMS *Rodney* and the HMS *Sheffield* in 1939.[79] Meanwhile in the United States, a device was developed in 1936 which could detect a plane at a distance up to 40 km (25 miles). An early radar system was fitted in the USS *Leary* in 1937, and from 1941 ground-based radar was operational in the US Navy. Civilian air traffic controllers came to rely on radar only from the 1950s, although already in pre-war years they had been in touch by radio with airborne pilots.

Colonies and Dominions

In 1899, the British government considered and discussed with Marconi the advantages of wireless telegraphy in imperial administration and links to the dominions. Connection of British-controlled areas throughout Africa as far apart as the Cape of Good Hope and Egypt would benefit from such a plan, while the island possessions in the Bay of Bengal could be linked by radio to the Indian subcontinent. Further discussions, conflicts and arguments between Marconi and the British took place in the years before the First World War, exploring the means to link London to British colonies and dominions, but the implementation of such ideas was long delayed.[80] Finally in 1924–1925, Marconi WT Company set up an Imperial Wireless Chain in Britain, Australia, Canada, India and South Africa.

Germany's initiative in developing wireless communication during the First World War reflected an earlier anxiety about the power Britain and France had gained in their mastery and control of cable networks to link their colonial empires.[81] Germany feared such global control would extend as wireless networks were developed.

Cabled messages from imperial capitals to colonial administrative centres, supplemented by the direct speech of the international telephone system, continued to be central to 20th-century colonialism. But telegraph links within a large colonial territory were limited, and the significant advantages of immediate wireless communication that emerged in wartime began to be seen for civilian administration in regions under colonial control, whose officials might be travelling and working far from urban centres supplied with telegraphy. Others – missionaries, traders, operators of widely distributed landholdings – could equally use radio communication to manage their operational activities.

For example, Britain's scattered South Pacific empire, largely administered from Fiji, was serviced by radio after an initial reluctance to see the advantages of the technology in colonial administration.[82] There were local tests from 1911 alongside legislative controls on wireless use. Visiting ships served to demonstrate the value of wireless communication across the Pacific oceanic region. The outbreak of First World War in 1914 had impacts that extended to competing powers in the South Pacific, and the disadvantages of delaying the provision of wireless communication became clear. The move to install radio equipment was prompt, with a radio telegraph station begun in September 1914 at Tulagi in the Solomon Islands.[83] Further development in the war years was accelerated in the 1920s, stimulated by the demands and needs of (European) planters and of missionaries, though not without some initial reluctance from the authorities.

Colonial administrations would also come to realise the value of public access to licensed and controlled national and regional radio stations broadcasting entertainment, news and propaganda. A territory whose boundaries were often artificial, drawn up after competition or agreement between European powers, could achieve greater identity and unity by the power of the radio reaching widely dispersed and diverse communities.

News and Propaganda

The wired telegraph allowed news to be transmitted fast across the world, exploited by the foundation of the Reuters news agency in 1851. But while the expansion of land and subterranean cables allowed the transmission of immediate information for newspapers to print, those at sea received no news. Marine captains were unaware of major breaking events of war and peace, and the wealthy who were travelling by ocean liners might not know of events that could affect their investments.

It was Marconi who saw the market opportunity and began to fill the gap. In 1899 while crossing the Atlantic on board the US liner SS *Saint Paul*, he collected news from the wireless station on the Isle of Wight and issued a printed summary of world events.[84] Daily news distributed from 1902 on the *Lusitania* inspired the major shipping companies to make a newspaper part of their passengers' daily experience.

The Changed Wireless World

The Wireless Press (otherwise called the Marconi Press Agency) from 1910 despatched the news each night to ocean liners, who could include it in the information printed daily for their passengers.[85]

When the First World War broke out, the speed, accuracy and value of news transmission were of crucial importance. Wired transmission could be subject to enemy sabotage. Britain maintained control of cables across the Atlantic after cutting all but one of those used by Germany to send messages to North America immediately on the outbreak of war in August 1914. German messages could be monitored on the remaining cable the Germans used.[86] The famous Zimmerman telegram sent in 1917 by Germany to Mexico via the United States was intercepted and decoded by the British, who were able to hide from Germany and the United States their code-breaking ability.

Germany used wireless telegraphy to spread propaganda, with a particular focus on reaching citizens of the United States, which remained neutral until April 1917. This was a crucial initiative and presaged a desire developed from mid-1916 to construct a world wireless network with stations in Mexico, South America, China and the East Indies, one which could bypass the risks associated with cabled communication in wartime. 'The world was no longer simply divided into land and sea. For Germans, the air had become the vital third dimension.'[87]

The British naval authorities, who had worked closely with Marconi, now authorised his Wireless Press to pick up German wireless messages, translate them and transmit in a context which exposed them as biased propaganda.

By 1918, the Wireless Press was also transmitting British official news to Paris and Madrid. During the later part of the war, it had expanded its news broadcasts to naval and commercial ships of both Allied and neutral counties. Wireless Press faced the challenge of how to distinguish propaganda from news.[88] Reuters maintained its image of independence even though it included information and perspectives biased in favour of the Allies: a British official described it as 'independent news agency of an objective character with propaganda secretly infused'.[89] The attraction of using an apparently objective wireless news service to serve political ends was explored in a post-war British plan for an 'Imperial News Service', which never eventuated.

From 1923, Reuters would broadcast news from Britain to other European countries using Morse code. It was by then the major source of international news for the British press.[90]

Reuters might be essential to print media, but its news supply service also became part of the most universally used application of wireless technology: the domestic radio receiver bought for entertainment in the home. In Britain, Reuters submitted news to the BBC from 1922 for evening broadcast only, to avoid competition with print newspapers. These continued to resist successfully the announcement of news as it happened, rather than a summary at the end of day. But with the nationwide General Strike of 1926, the BBC began to broadcast through the day news that its own staff had collected, and from 1929 it used Reuters as an agency providing news that it would itself edit and use as it wished.[91]

Entertainment Broadcasting

While wireless technology may have begun with services to maritime users, and proved its potential to governments in war and peace, its wider impact came with home entertainment. Commercial enterprises and governments alike engaged with the potential of wireless communication to reach out to domestic listeners. The years after the First World War saw a rapid spread in country after country of services which the average family could receive at home on 'the wireless' (as in British usage) or 'the radio' (as in the United States), tuning in to a range of alternative programmes when they were 'broadcast' from faraway transmitters.

The potential was first indicated at the end of 1906 when Reginald Fessenden was able to transmit from Massachusetts sounds of music and voice (though to an unconfirmed audience). Further experiments continued, though initially more slowly than the other applications previously described. The First World War put entertainment radio on the back burner, although in the middle of the war German troops received music from a transmitter in Cologne.[92]

The Marconi Company had led many of the initiatives but held back on entertainment radio. One of their employees, the Russian-born David Sarnoff, made the prescient proposal in 1916: 'I have in mind a plan of development which would make radio a "household utility" in the same sense as the piano or phonograph. The idea is to bring music into the house by wireless.'[93]

The commercial potential of domestic radio was developed not by the Marconi Company but by other organisations such as the Radio Corporation of America (RCA) owned by the General Electric Corporation, which provided David Sarnoff with an opportunity to take his ideas forward. Westinghouse became a major owner of RCA and under

The Changed Wireless World

government licence in 1920 it began public broadcasting with KDKA in Pittsburgh, starting with just one hour each evening.[94] The commercial value to Westinghouse was initially to encourage the sale of domestic radios; in 1921, Westinghouse offered the Aeriola Jr for $25.50, and later that year the Aeriola Sr (Figure 37). Radio parts were also major sales items, as the majority of early receivers were constructed at home. In a speech by Herbert Hoover (then Commerce Secretary) in 1922, he noted that ownership of radio sets had moved from 50,000 to 600,000 in one year.[95] By the 1930s, portable radios were available for purchase.

Commercial sponsorship on the radio began as early as 1922 with New York City station WEAF. In the United States, advertising on radio programmes soon became an important source of income and profit, allowing and encouraging the development of new radio stations whose owners had no receivers to sell. By 1924, there were over 3 million radio receivers owned in the United States (by one in ten households), and by the end of that year there were an estimated 530 radio stations.[96] In rich areas or

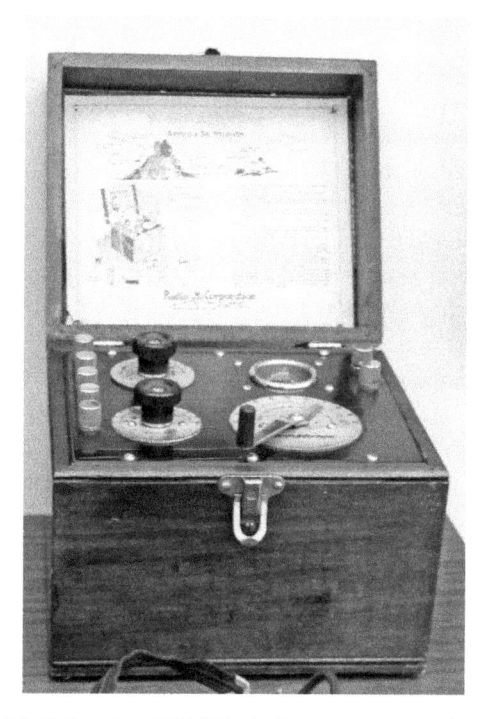

FIGURE 37 Aeriola Sr Receiver, RCA Westinghouse, 1921–1922 radio (Fourandsixty, CC BY-SA 4.0, via Wikimedia Commons).

poor, in prosperous times or (as in the 1930s depression) more challenging times, radio would provide free access to multiple forms of entertainment. Music, talk, drama, news and sport (a boxing match broadcast in 1921 being a pioneering venture) were freely available within the home.

Radio developed also as an invaluable source of information as well as entertainment. Accessing news is now considered a major contribution of broadcasting, but news was not a significant feature of US radio in its early years. Once established, a survey suggested that one of the main reasons listeners tuned in to news broadcasts for the weather forecasts they included.[97] In 1922, some 20 of the 36 commercial radio stations were licensed to transmit the weather forecasts supplied by the Weather Bureau, but by the following year such licences were granted to all 140 operating radio stations.

Across rural areas of America, farmers could receive weather warnings as well as information on market prices for their produce.[98] Also, not insignificantly, families on isolated farms could participate more readily in multiple aspects of wider society: skills, music and learning.

Different countries followed different paths. The Netherlands laid claim to the first public broadcast from 1919. The Argentine Republic had an early history with broadcasts from 1920, as in the United States. Australia too began with commercial radio, as 2BL and 2FC went to air in Sydney in 1923. By that year, radio stations had been launched in countries as widely distributed as Brazil, Canada, Chile, China, Finland, Mexico and New Zealand.

The United Kingdom took a different path from the United States. Marconi had run experimental sound broadcasts in 1920 from Chelmsford. British enthusiasts found they could hear other broadcasts originating in Holland and France. The government owned British Broadcasting Company had its own experimental broadcast in November 1922.[99] It operated formally from 1923 (it was renamed Corporation in 1927) and eschewed advertising, with broadcasting seen as essentially a public service, funded by radio listeners' licence fees, with a rather limited initial view of what was considered appropriate entertainment. By December 1923 some 595,000 licenses had been issued; by June 1927 this had risen to just under 3 million, reaching 10 million by the end of the Second World War.

Britain indeed remained free from legal commercial radio until 1973, requiring those of us seeking a wider range of music to tune in to Radio Luxembourg or offshore 'pirate radio'. But the British pattern did encourage imaginative programming funded by the public purse via the radio

The Changed Wireless World

licence. The BBC had its own orchestras, in popular music from 1928 and classical music from 1934. The first radio drama, A *Comedy of Danger* by Richard Hughes, was broadcast in 1924. Many elements of cultural and intellectual life in the 1920s and especially in the 1930s can be credited to radio influence, with talks, daily short stories and more.

The growth of radio in Russia paralleled the development of Bolshevik rule. Early demonstrations of radio in 1919 and 1920 were followed by a loudspeaker broadcast in Moscow in 1921 and full broadcasting by state broadcaster Radioperedacha from 1924. However, much of the transmission was initially by wired networks, with public loudspeakers available in many cities, and home-made crystal sets with limited range.[100] By 1928, there were fifty-nine stations in the Soviet Union, and radio could reach distant corners of the territory.[101]

Television

It is striking how soon after the establishment of sound broadcasting were the first moves to extend this to visual broadcasting, what would develop as television. Again, much experimental work by widely distributed researchers led to what would become a commercial reality. The developments associated with Scotsman John Logie Baird (1888–1946) integrated many of these innovations in a working system demonstrated publicly in 1926, even extending to colour images in 1928.[102]

Marconi that year gave his support to the potential, stating 'I believe that television is destined to become the greatest force in the world – I think it will have more influence over the lives of individuals than any other force'.[103]

In Britain, the BBC committed to experimental television broadcasts (after initial reluctance) in 1929, and by the following year a small number of owners of television receivers could observe the output of these experiments. Occasional broadcasts projected onto large public screens added to the audience. Fewer than 1,000 receiver sets had been sold by the beginning of 1931.[104] The 1930s proved a difficult economic time for further public investment and saw difficult relations between the BBC, Baird and other interested parties, despite the confidence of radio manufacturers in the future of television.

Eventually on 2 November 1936, the BBC officially began regular television broadcasting, from the 1873 building Alexandra Palace in North London, which had been adapted to provide studios suitable to a range of programmes (Figure 38). Early broadcasts included music, ballet, talks and film excerpts as well as demonstrations of sport from outside the building.[105]

A broadcast of the 1937 Coronation extended the appeal and power of television. Take-up of television sets was slow because of their expense (though falling to under £60 in 1937), the quality and distance from the transmitter, as well as the limited number of broadcast hours. Numbers of sets reached a little over 2,000 in 1937. By August 1939, there were 20–25,000 sets in use, but then the outbreak of the Second World War put different priorities on the public purse, and television broadcasts were stopped, not to resume until 1946.

Despite (and perhaps partly because of) the success of commercial radio in the United States funded by advertising revenue, television was slower to enter that market. There had been successful experiments as early as 1928, continuing with some university and private broadcasters in the 1930s, and

FIGURE 38 BBC Television camera with cable link to Alexandra Palace, 1938 (photo credit: Chronicle/Alamy Stock Photo).

occasional public displays, but the poor visual quality discouraged further investment. Work by Vladimir Zworykin on cathode ray technology, which gained support from the RCA, led to developments with promise for the future.[106] The cathode ray tube had its origins in much earlier work by German scientist Karl Ferdinand Braun from 1897.

By 1939, RCA under David Sarnoff announced regular broadcasting by NBC based on cathode ray tube technology, marked with a special programme based at the New York World's Fair.[107] Purchase of a television set required the incentive of enough worthwhile programmes to justify its cost ($200 for a small tabletop set, $600 for a 12-inch version); investment in programming required enough viewers to secure paying advertisers. In 1939, there was enough confidence that this model would work. As many as 3,000 households or more public venues acquired a set, and regular commercial television broadcasting grew without pausing for the Second World War. By 1946, there were still only 6,000 sets in use, before a massive growth to an estimated 13 million sets by 1951.

While other countries had been stimulated to undertake their own experimentation with television in the years before the Second World War (with notable innovations by France), it would take until the early 1950s for consumer television to spread more broadly.

National Unity in Language and Culture

It is impossible to underestimate the power of wireless communication, in the form of broadcast radio (and then television), in creating or strengthening community identities in numerous parts of the world. Consciously and unconsciously, governments, ruling elites and cultural leaders found in wireless technology an instrument of influence and integration, as access to domestic radio expanded. Later, television would play much of the same role. This echoed the effects achieved earlier by the written word and then by the power of print.

Most nation states of past and present have boundaries which are artificial, shifting and changing through history. Family loyalties may lie first in their community, or their language group, or religious affiliation, or the ethnic identity they claim, or the historical groups (or nations) they identify as their heritage. A national broadcaster can bring different groups together, sharing a common experience, listening to a common language or even accent, hearing the same news (and propaganda) and identifying with a wider world through ready access in the home to culture, however we define that.

We noted previously the use of radio by the Bolshevik rulers who sought to develop a single nation of the Soviet Union where once Russia had an empire of diverse ethnicities and loyalties spread across Eurasia. The history of 20th-century central and eastern Europe saw boundaries shifting, with some new nations emerging and some dissolving. New states in the Middle East developed with the end of the Ottoman Empire.

Colonial territories had artificial boundaries set or agreed by European powers in the 19th century, with some alterations following the First World War. National boundaries linked previous separate and even hostile ethnic and language groups, while dividing other communities with map-drawn borders which became formalised in time. Island territories in the Caribbean and more widely in the Pacific were grouped and defined as entities by the powers which ruled them (eventually to emerge as new nation states in the era of decolonisation).

The language of the colonial powers might be conveyed not only in the legal and educational system but extended to wireless broadcasting. Alternately a colonial (or post-colonial) government might seek to affirm its power by broadcasting in a set of separate community languages.

In July 1927, the Bombay (Mumbai) station of the Indian Broadcasting Company, licensed by the government, transmitted the Viceroy's voice across India, seventy years after the rebellion that had shown the value to the British of the electric telegraph.[108] A decade of more experimental exercises in Indian radio preceded this launch of English-language broadcasts. In a context of the emerging strength of the nationalist movement, radio should have been of heightened potential value to the colonial authorities, but take-up of domestic receivers began slowly. With only a few residents signed up, the radio company failed, and in 1930 the government themselves took over broadcasting from the company. Through the 1930s, numbers of radio subscribers grew only slowly, and it would take until the emergence of transistor radios in the 1960s that radio became universal across India.

Broadcasting in a national language could reinforce the use of that language. It could also prioritise a pronunciation – 'BBC English' being the British pronunciation associated with an educated southern elite, although very different from the common dialect of the capital. The BBC had an 'Advisory Committee on Spoken English' from 1926. This advised both on the use of words and on their pronunciation in programmes; playwright George Bernard Shaw served thirteen years on this group, including seven as chair.[109] As the British government-funded BBC developed a programme of world broadcasting, it had significant influence on listeners worldwide for whom English served as a second language.

Unifying a nation or language group had immediate political rewards as well as the more abstract effects of cultural integration. Radio in Germany had modest beginnings in 1923, but that would change. The rapid rise of Nazism gained much strength from using radio for propaganda. 'By 1933, the wireless had already become the most powerful instrument of political influence. The Nazi leaders were able to manipulate their broadcasting system in a way not previously imagined in a democratic state.'[110] Radio was the Nazi party's weapon in gaining support and asserting authority in pre-war Germany, under the arch-propagandist Paul Josef Goebbels, who could boast in 1933 that 'broadcasting is now totally in the hands of the state'. But of course that was no different from Britain or Russia, and a contrast with the commercial radio of the United States.

British prime minister Neville Chamberlain's announcement of the outbreak of war could reach 32 million listeners on 3 September 1939, by when three-quarters of the population had wireless access, and wartime news further increased this demand.

Forging national identity by means of broadcasting could also mean a central broadcasting authority using regional languages as a means of resisting separatism.[111] These were often reactive: the BBC had regional radio programmes in English to supplement the national ones from the early years of broadcasting, but only much later in Welsh (1976) and Scots Gaelic (1985). By contrast, the Irish Free State established a radio station from 1933 which used mainly English rather than the Irish Erse language.

Connecting the World

The role of wireless in emphasising nationalisms, supplying wartime propaganda and providing new military technologies can overshadow its contribution to internationalism and international relations. Maritime safety had relied on cooperation between nations. This was achieved by the international acceptance of Morse code and the collaboration on radio frequencies advanced by regular international conferences of maritime nations to agree common procedures.

Likewise, the sharing of weather information was essential to maritime safety, even though the distribution of weather warnings to ships was the responsibility of individual nations.

We think of time as an absolute, but the measurement and correlation of time had been an approximation based on the technology of clocks, watches and maritime chronometers. Since radio communication made it possible to send a time signal over land and sea, the international use of radio allowed the correlation of such time signals and the development of what we now take for granted: an absolute measure of time wherever we are in the world. From 1905, radio became the basis for time signals and time coordination.

News agencies and news journalists could now transmit their information instantly across the globe by wireless means as well as cable, and in turn radio stations could transmit world, national and local news to their widely distributed audiences. Information on world events could now cross the globe soon after these happened. Immediacy was not just limited to the exchange of cables between and within governments, or available to the business and finance communities. Distant events could be brought readily into the home, even if awareness of a single world failed to prevent nationalisms and wars, ethnic and religious conflicts.

Criminal Apprehension

Just as the early railside telegraph had proved its worth in 1844–1845 in the apprehension of British criminal figures, so wireless communication changed aspects of policing and crime control.

In 1902, there was a hold-up in a bar on Santa Catalina Island off the coast of California, and a radio message to the mainland allowed the robbers to be caught.[112] Another early proof of radio's value in combating criminals was in 1910, when American Hawley Crippen and lover Ethel Neave fled England by the transatlantic vessel *Montrose* after Crippen had murdered his wife. The ship's captain, knowing the police were after Crippen, identified him and was able to radio this information back to London, allowing a detective to follow by a faster vessel and ensure their arrest when they landed in Canada.

The potential of radio links by police to their base (or to each other) would provide a major contribution to crime control and public safety. This began with a maritime use: in June 1916, the New York Police Department established a private coastal base station to communicate with police boats in the New York harbour.[113] In 1923, Pennsylvania linked major police stations by radio telegraphy. The Detroit police force experimented with radio through the 1920s, and in 1929 was operating a short-wave radio system for their police force. By 1937, some 226 US cities had two-way radios to link their police.[114] This contrasted with British police forces, who were far slower to incorporate radio communications into their work.

Financial Operations

Speedy information transmitted immediately by a wireless radio system might be a matter of life and death in a maritime context, or to land- or air-

The Changed Wireless World

based military units. It might be of value in providing weather warnings to those on land, and news broadcasts would become of wide interest as distributed listeners shared in information on the day's events reported by wire services through their favourite radio channel. But to those in the competitive worlds of finance, quick access to relevant news and quick reporting of movements in the prices of stocks, currencies or raw materials would be seen as priorities.

The provision of a paper 'ticker' to report stock market prices in real time began in the New York Stock Exchange in 1867, available also in the offices of major brokerage offices.[115] Two revolving wheels powered by a battery printed off the letters of a security and the price information. From 1870, these were also printing off the volume traded. Slight variants of this original approach (including initiatives by Thomas Edison) sought to avoid the clash with the patent of originator Edward A. Calahan. With the extension of the method, information was despatched to brokers and investors outside of the New York hub, with perhaps more than 1,000 outlets by 1900 (though the numbers are disputed).[116] Ticker tape – in use until the 1960s in investment houses – became such a familiar concept that it was associated with the acclamation of heroes in 'ticker tape parades' in central New York.

Stock prices and other financial information were provided in timely fashion by agencies. Established in 1872, Extel (the Exchange Telegraph Company) gained the exclusive position to report from the London Stock Exchange and set up ticker-tape machines in Britain.

Radio allowed a broader speedy distribution of financial information. Reuters was a major provider of financial news, which was included with wider news in the media of the day as radio broadcasting developed. A notable landmark was the Wall Street Crash – the suddenness of the declining values of the New York stock market in October 1929. News agencies reported the changing stock values as breaking news, picked up by broadcast radio and then by daily newspapers. Reuters reported:

> New York Tuesday [29 October]: Wall Street has never witnessed such a wild opening as this morning. Practically all the leading stocks opened with initial sales of from 10,000–15,000 shares, with average decline of up to 10 points ... the closing gong at three o'clock ended a day of greater pandemonium, apprehension and general uncertainty than any broker has previously experienced.[117]

The news, of course, both reported and stimulated the panic selling. Meanwhile the ticker-tape machines ran longer hours than they had ever been scheduled to run. Later, when Britain moved off the gold standard for its currency in September 1931, it was Reuters radio wires that alerted the world.

FIGURE 39 Wireless operator recruitment during the First World War (photo credit: Chronicle/Alamy Stock Photo).

New Careers

Unsurprisingly, uses of the new technologies of wireless communication required new careers – many of which could not readily be adapted from existing roles.

The specialist nature of radio communication at sea needed training and skills quite different from those of the traditional navigator. Radio operators needed both to send and to translate messages in Morse code but also to have basic knowledge of maintaining equipment. The Marconi Company opened a School for Wireless Operators in Britain in 1902.[118] This provided an accelerated training (as little as just one month), indicating both the urgency of the new demand and the potential for employment of graduates of the course. Other training centres developed (Figure 39); in 1919, a college in Cardiff could advertise 'Every student of ours was offered a berth the same week he qualified'.[119] Naval operators, of course, needed to be enlisted members of the armed forces. The US Navy would train wireless officers in seven to ten weeks.[120]

Amateur radio enthusiasts might have developed their passion without expecting it to become a career choice, but when the First World War broke out, those in Britain could find themselves playing an unexpected role in serving the army.

FIGURE 40 Radio assembly work, Philadelphia, 1925 (photo credit: Granger – Historical Picture Archive/Alamy Stock Photo).

Early entertainment radio in the home stimulated many to bypass the expense of buying a commercial radio receiver and to construct their own crystal set. In time, mass production of radio sets reduced the cost in the home. Radio assembly work in factory assembly lines proved an employment opportunity especially for female workers (Figure 40).

The rapid development of broadcasting in the 1920s engaged established entertainers from outside the industry together with technicians, producers and commercial managers focused on the potential of the new technology. Musicians might alternate between studio and public performances, but the spoken word allowed writers of radio plays and those delivering radio talks to thrive alongside news readers and continuity staff.

Low-Power Devices

Human imagination and commercial imperatives combined to explore and secure numerous other applications of wireless radio. Low-power devices with radio waves communicating over just a short distance made for many possibilities for experiment, taken up by amateur enthusiasts long before commercial manufacturers developed products to sell into the market.

Model boats on a lake, model cars on a track and model aircraft could be moved by radio controllers. Tesla's 1898 patent was for a remote-control system which he demonstrated in controlling a model boat.[121] Separate radio waves were transmitted to adjust the rudder and to set the speed and direction of the propeller. This was a dramatic demonstration of the power of radio controllers, but the cost limited the pace of expansion before the 1960s.

Metal detectors, used by hobbyists to search for buried antiquities and lost valuables alike, had an earlier origin. Initial examples in the 1920s preceded the more ambitious development of mine detectors in the Second World War. Consumer hobbyist metal detectors were commercialised from the 1950s onwards. The ability to open a garage door remotely from outside was developed by individual enthusiasts in the 1930s and first commercialised in the United States in 1954. Baby monitors were first sold in 1938 in the United States by the Zenith Radio Corporation. Hobbyists were responsible for developing the first wireless microphones, but they became commercially available only by the end of the 1950s, a significant moment in the history of popular music.

The Internet: Cables and Wi-Fi

It is worth noting that one of the most dramatic developments in the flow of information, the Internet, involved the ability of *cabled* telegraphy to convey data. The services developed within the Internet have come to include email messaging, the world wide web, and much more. The conventional account places the origins of the Internet in the US Defense Department in the 1960s: as with the naval support for early radio communication (and so much else in scientific progress), military and defence budgets, needs and support underlay the innovation. Fundamental to its value was the ability to send significant blocks of information speedily between distant computers, and this utilised physical cables rather than radio waves.

Internet use and applications spread to commercial, government and university research before a development of public access from 1989, growing fast in the 1990s, marked the beginning of a transformation. But this was still a wired process, as businesses then homes linked their own computer system to the Internet through telephone lines, with the subsequent massive growth of email and development of sites accessed on the world wide web.

Experimental developments to create a wireless local area network soon followed, and adoption of Wi-Fi technology spread through institutions and private homes from the early 2000s. Yet the efficiency of Wi-Fi is dependent on the speed and efficiency of the networks of fibre-optic cables, a reliance on the development that preceded the taming of radio waves. Our current expectations and emphasis and reliance on fast broadband Wi-Fi access in home and office requires cabled connection to a building to reach a modem.

Radio Telephony

Perhaps the largest visible impact of wireless communication today is the mobile phone (cell phone). Wired telephones from the 1880s (only transatlantic in the 1920s) provided apparently adequate means of voice communication for those on land, leaving priorities for radio voice communication to those at sea or in military positions.

The idea of extending radio telegraphy stimulated debates; one concern voiced in discussions of 1920 focused on the issue of privacy. A wired telephone between two linked speakers was private; a conversation on the open radio waves was at risk of being overheard by others.

In 1927, the General Electric Company demonstrated a system of short distance radio telegraphy on trains for the guard and driver to communicate. Truck drivers, initially in the United States, adopted CB (Citizens Band) radio in their cabs from the 1940s onwards, and CB radio developed for other uses.

It was the experience of land, sea and air forces in the Second World War which served to emphasise the value of two-way radio. Indeed, radio communication was a key to life and death in military manoeuvres, with each platoon, even each tank, in contact with command centres by radio. The technology continued to develop rapidly through the stages of the war.

Today's ubiquitous mobile phones and mobile networks reflect a complex sequence of development, with practical applications created through the 1990s. Citizens became no longer reliant on the availability of a working wired telephone. In substantial areas of the world, for residents of rural communities where electrical supply was patchy or unavailable and wired telephones absent or faulty, the mobile phone now provided (though at a cost) a democratic access to communication. Families could be linked worldwide, and workers could seek jobs far away. With the development of smartphone technology, financial transactions put banking (whether payments or the transfer of funds between family members) in the hands of villagers living far from traditional banks.

Today's smartphone is just another stage in the development of wireless communication, though its combination of numerous applications of radio waves in one instrument serves to emphasise the revolutionary impact of a technology that began only around the beginning of the last century. The smartphone includes non-wireless functions such as a compass, camera and video recorder. Using radio technology gives us accurate time and provides weather forecasts, while its maps (replacing atlases) use satellite-based radio transmissions to show us exactly where we are and direct us to our destination. Mobile internet access links us to worlds of information and provides us with words, music, images and film wherever we are. We transmit photographs and documents by wireless transmission to cloud storage.

The smartphone gives us wireless voice telephony, video telephony, text messaging, email and social media. It allows us to keep in touch with an individual family member, to engage in courtship at a distance, to address a broad group of friends simultaneously or to publish our thoughts, images, prejudices and delusions to as wide a world as is willing to see them, through expanding commercially owned and controlled social media platforms. In turn, varying with location, commercial and government organisations can turn social media into tools of control, monitoring, influence, income, manipulation and the construction of databases. Social media via the smartphone presents freedom to communicate and an assumption of the democratisation of communication, but as in the revolutions of writing and printing, there is an imbalance of power underlying the impact of the transition.

Conclusion

The perception that radio waves could be tamed to meet human needs led to numerous experiments which developed into commercially viable technologies – technologies that changed human lives in one, two, three generations. New adaptations would emerge. Space exploration combined radio controls from earth with information sent back by satellite and space modules. Worldwide access to information and messaging would benefit from Wi-Fi linked to the Internet. Numerous scientific studies would build on the potential of wireless technology – from GPS systems recording archaeological finds to biological studies using tracking devices to monitor the movements of wildlife, as begun in the 1960s. An extension of this meant prisoners released on parole or bail could be required to wear electronic tags whose radio transmissions signalled their whereabouts.

The speed with which human lives have been transformed by the wireless revolution is hard to comprehend; harder, perhaps, because we are the middle of the process of rapid change. If we look back at the multiple impacts of fire, horse domestication, writing or the printed word, we can see changes as a fulcrum distinguishing what followed from what preceded those innovations. When our own lives have been in the middle of a transformation process, it is more challenging to make sense of it all.

The wireless world which has developed over little more than a century marked an end to forms of communication which had remained similar and limited for five millennia of urban civilisation, twelve millennia of farming societies, over 200 millennia of our modern human species or two and a half million years of our stone tool-making ancestors.

If we stand back a little from history, it is easy to acknowledge the differences in personal human life, interpersonal interaction and the organisation of society between the late 19th century and today's wireless era. Linking to anywhere in the world while on the move, transferring money from a distant city to a remote village home, hearing world news (even watching it unfold in real time) mark changes which no one in the 19th century could predict.

The full potential of harnessing radio waves has not been reached, whether this lies in the further development of space exploration in the solar system or in integrated communication to further transform administration and business. There will be new unanticipated impacts, but there will also be new technologies currently unimagined which will be just as transformative as those described in this book.

CHAPTER 7

Innovation, Progress and Presentism

In the previous five chapters, I selected and discussed innovations which can be said to have changed the history of humanity. Origins and developments in one continent would in time spread worldwide to reflect the advantages of a new facility. Discoveries and inventions may be in extended stages (as with the uses of the domesticated horse) or progress more quickly. One key is how readily new ideas could be communicated, and communication is itself core to several of the themes of the book. The inventions of writing, the printing press and use of radio waves were major steps to improve communication across space, and the ridden horse also sped up links between communities.

I have sought to demonstrate and emphasise how many and diverse can be the impacts of a single change in technology or economy. An innovation can reflect a social need but can also act as catalyst for changes. Such changes can be in personal and family experience, social life, economic patterns, power and authority, belief systems and much more – even biology. Readers of this book might question some of my suggestions of the short- or longer-term impacts of the individual innovations discussed; or they might add their own proposals to the catalogue.

This stimulates some broader questions: what distinguishes some innovations from others, since history (and prehistory) only make sense when we describe what changed: where, when and how. More broad are questions on what we might define as human 'progress'. Is there a linear development implied by innovation, or does progress in historical terms mean something quite different? And can we ever agree about this?

Are we so biased by the experience of our own lifetimes that we are blinded to the long-time scale and nature of human or hominin existence. Is presentism a prejudice which we can avoid? And when we are told that we are now entering a new geological age, the question of own species' significance looms large.

Innovation

What marks out some innovations, such as those I have featured in this book, is that they separate in both time and community those who had the new skill or technology, and those who did not. The advantages of being able to create and maintain fire, the ability to ride a horse or haul a wheeled vehicle are such that, once gained, they would not readily be abandoned. They provided special powers to communities with those skills. When urban societies developed or acquired the innovations of writing, and later printing and finally wireless communication using radio waves, the potential these changes provided were substantial. There was advantage within society to those who could access such new media. A small minority of the societies of the ancient Middle East would be literate: the elite and their scribes. A larger group but still a minority had access to the printed word during the 15th-century development of the printing press. Radio communication spread much more rapidly by its very nature, but control of communications privileged state and commercial bodies, arguably until the Internet began to challenge such control.

What links these specific developments is their long-term impact, even into the modern world. We may have become a little less reliant on fire in our daily lives, and the generation of electric power using sustainable means rather than coal- or oil-fired powered stations will further decrease our reliance on fire. While the horse may be less of a daily experience, the wheeled vehicle is sure to remain part of human society. Despite the impact and potential of electronic publishing on the printed word, print dominates children's reading and continues strong in books published for adult readers too, even if the daily newspaper and the academic journal are disappearing in physical form. As we continue scientific developments and space exploration, radio waves remain core to our future.

The last point foregrounds one feature of many long-term innovations: that they can reduce social distance. Fire itself provided small-scale forager communities with a hearth as a centre for a social group. The horse enabled people to travel further and faster than on foot and, once linked to a wheeled vehicle, made it possible to transport goods beyond the time and spatial limits of haulage by humans or pack animals. The movement of goods and people also meant the movement of ideas. So the wheel and horse shrank the geographical and social worlds.

Innovation 237

Means of communication also overcome the tyranny of distance. A written message sent by courier, like a monumental declaration by a ruler or indeed a graffito on a rock along a trade route, meant that writing had bypassed the limits of face-to-face human speech. The printed word took this much further, as a single text could travel throughout and beyond Europe and be read a week, a year or even centuries after manufacturing. In the Protestant revolution, print could bring the words of revolutionary new preachers across the continent and bring the texts of Old and New Testament in vernacular translations at the same time, with classics of the ancient world also readily available. Wireless communication meant the world had shrunk yet again, with the ability now to watch in real time an event unfolding at the other side of the globe.

In Chapter 1, I observed that the topics chosen for this book reflected origins in different parts of the world. But their significance and potential were such they enabled humans to create new history in regions far from that of their origin. If the Neanderthals of Europe could perhaps only use fire from natural sources, the ability of anatomically modern humans to create fire as they spread outward from Africa allowed access to areas and seasons which might otherwise have been inaccessible, including their migration from Asia into North America. The power of the horse would in time allow fast movement of warriors from Central Asia to conquer vast regions from China to Europe, while the chariot and cavalry had earlier come to dominate much of the world of competing empires of the Middle East. Europeans arriving in the Americas with their horses changed the trajectory of the whole New World. Printing with metal type began in Europe (with a print shop opening in Mexico by 1540) and was timed to report the Age of Discovery to a newly expanded audience of readers. The development of radio technology to serve entertainment and communication worldwide owes much to the commercial imperatives of North America. Innovation can break down the barriers of geography and continues to do so with space exploration.

The subjects chosen for this book are examples of human developments with long-term and widespread impacts. But there are many other examples. The domestication of food plants in both Old and New Worlds meant not just greater control over food resources but the need for a settled life to plant, maintain, harvest and process these foods. The stages of domestication were gradual and occurred in separate areas of the world, with varying implications for culture and settlement patterns, but led to significant increases in density of human settlement, the first localised population explosions. The innovation of village life is in itself an innovation, but also a gradual process. Nevertheless, plant domestication is a different process from that of domesticating food animals. The pastoralist, like the hunter, moves with the seasons, perhaps grazing animals in the highland in summer, before returning to lowland in winter. It is the farmer of plant foods who must have a long and stable base.

More dramatic, perhaps was the Columbian exchange after the 15th century. European crops came to dominate many colonial possessions in the New World (with subsequent social impacts including, of course, the transatlantic slave trade) while food native to the Americas changed the economic lives of many societies in Europe, Africa and Asia.

My chapter on writing was set against the background in which urban civilisations began, and the urban revolution involves several related, though not simultaneous, developments. But every city (as a trading, specialist, temple and state centre) requires a large hinterland of productive agricultural settlement to generate the surplus absorbed and exchanged within the urban setting.

We can no longer see monumental architecture as a singular innovation, not least because excavations at the sites such as that of Göbekli Tepe in south-eastern Turkey have revealed dramatic construction by pre-pottery communities not reliant on domesticated plants or animals.

Technological developments often dominate our concept of human prehistory and history. Pottery is such an invention: the discovery that clay could be moulded and hardened by heat to provide a permanent receptacle for liquids and food and for use in cookery. Pottery was invented in widely separate areas but is not a universal, and its widespread association with the development of domestic plants is not inevitable (the Jomon culture of Japan being pottery-making foragers).

Inventions in metallurgy were also key moments in human prehistory and history. The acquisition of an ability to extract copper from ores and make it into tools and weapons marked a major new technology in societies, and creating alloys to produce stronger items of bronze marked a true transition. While bronze might have been a valuable and often even elite possession, it could be melted and reused. The introduction of iron again distinguished societies with this new facility, before further developments such as steel kept metallurgy at the fore of human achievement.

While pottery and levels of metallurgy are distinguishing features notably used to categorise prehistory, especially in the Old World, modern history is dominated by key innovations which have been so transformative that their impacts on society, economics and politics cannot be underestimated. The harnessing of steam power effectively marks the beginning of the industrial

Innovation

era. Aircraft (and spacecraft) changed our perception of distance in the 20th century. The provision of electricity, to supplant steam in industry, gas in the street, fire and more in the home, would be as dramatic an innovation as any in human history. Without electricity, there would be no computers, and the computer age is now so embedded in our lives that we inevitably mark it as a development with just as significant a role in changing human society.

For a technological innovation to become the basis of a social innovation, the timing has to be right. As often noted, supposedly new technologies can often be traced back to an early proposal which did not seem to fit the needs of the time: Leonardo da Vinci and the helicopter or early 19th-century experiments with computers could be examples, or anomalous early prehistoric examples of ground stone.

Innovation in ideas is generally a more gradual process. Religious changes are typically gradual: Judaism emerged stage by stage, while Christianity had small numbers of followers until the 4th century AD.[1] By contrast, Islam spread rapidly in the Arabian Peninsula in the 7th century AD, but its broader early influence can be attributed to the political power of invading Arab armies in western Asia and North Africa, and even then the impact on local economy, culture and society was not immediate. Historical narratives are dominated by changes in rule and empires. While material culture and social organisation can be transformed dramatically by conquests, such changes may not outlive the cycle of dominant rule.

These innovations often had a specific beginning stage which had lasting impact across multiple aspects of society. That makes them different from changes which developed and evolved gradually over time, even if these might be fundamental in human history. The most obvious of these is language. Theories of the date by which we can speak of language vary, influenced in part by the study of the human cranium and jaw in fossil hominins and in part by assumptions about the importance of interaction between human groups and their need to communicate for food gathering, settlement and defensive strategies.[2] So whether we describe language as the sounds made by our hominin ancestors and relatives or assume it to be a feature just of *Homo sapiens*, its development would be slow.

Similarly clothing: an important aspect of evolution and adaptation, but one not essential to human life, even in areas of extreme seasonal cold. Adoption of clothing was a gradual process influenced by perceived need, available materials and technical skills to use those materials.[3]

Equally slow in its evolution (and even more open to untestable hypotheses) is the development of world views which we might include under the term 'religion'; and linked closely to those, ritual behaviours. We can periodically see evidence of such activities in prehistoric settlements in the arrangement of space. Activity reflecting ideas beyond the strictly functional is most visible with the emergence of art – a category which spans from scratches and ochre marking on stone plaquettes to rock engravings on walls and paintings in caves. Our understanding of the early development of 'art' is limited by the non-survival of organic material which may have been carved or decorated as objects of art or ritual.

Our knowledge of the beginnings of watercraft is also limited. We know that early *Homo sapiens* crossed from the Indonesian Sunda islands to New Guinea, probably more 60,000 years ago, and it seems highly likely that humans had earlier taken the water journey into southern Arabia across the Bab el-Mandab from East Africa. But movement of anatomically modern humans through Africa, Asia and Europe was mainly by land and could exploit terrestrial, lakeside and coastal resources without the consistent need for the knowledge of watercraft. We can interpret this as a development based on the specific demands of time and place.

Our study of prehistory, especially of early foragers, was traditionally based on classification of their stone-tool technology and major changes in method. Pebble tools developed to bifaces; later, well-created tools from a prepared core gave greater potential and range; and later still, punch-struck blades were developed and sometimes used in composite tools. Small, carefully prepared stone elements provided more complex stone tools, and during this development tools and weapons were set into handles. Even the development of agriculture was described as the 'Neolithic Revolution' in the Middle East and Europe, signifying a change to the use of polished stone tools. So there were discontinuities and developments even if we do not speak directly of 'inventions' in stone-tool technology.

If we were to consider the most successful innovation in the history of humankind and our hominin relatives, it is too early to say that it was electricity or the computer, whose longevity remains unproven. It is more convincing to put the bifacial stone handaxe in that role. From around 1.6 million years ago and still to be found in use as late as 200,000 years ago, the symmetrical stone handaxe in the Lower Palaeolithic Acheulean cultures of Africa, Europe and western Asia was a core part of the technology of our *Homo erectus* ancestors (Figure 41). At its most perfect when created from flint or chert, the handaxe combined a pear-shaped (or teardrop) profile with roughly parallel edges and carefully flaked surfaces.

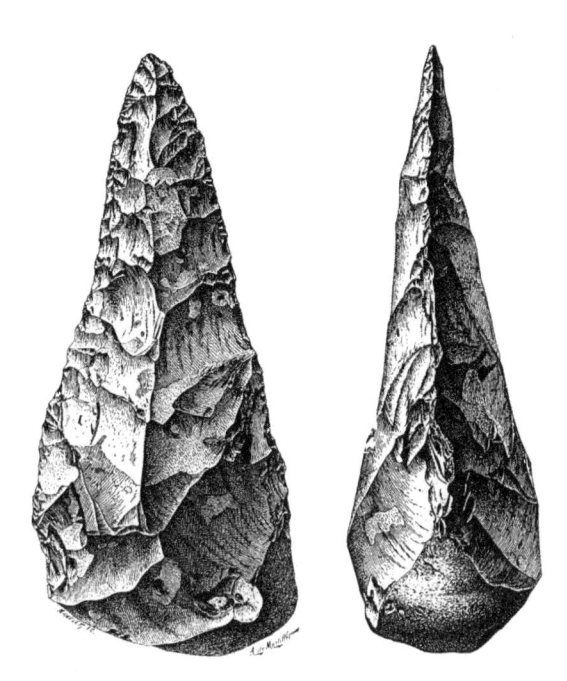

FIGURE 41 Acheulean bifacial flint handaxes (A. de Mortillet, public domain, via Wikimedia Commons).

Usually able to fit into one hand, it thus served as a multipurpose tool: sharp point, broad base, sharp edges, suitable for uses ranging from butchering animals for meat to cutting wood and processing plant foods. It doubtless also served to demonstrate the competence of the maker, perhaps important in mate selection.[4] The skill involved in making the handaxe (a time-consuming process compared to other stone or wooden tools) conveyed the biological message of 'good genes', identifying the talented males (given the traditional assumption that this was a gender-specific activity).

We know that the users of stone handaxes were also using tools of organic materials – wood, bone, shell – but the survival of these is much slighter than indestructible stone. Tools are part of the means by which groups fit into their environment, use its potential for survival and where possible modify it. Ubiquitous contemporary discussions focus on the relationship between our species and 'the environment'. Some of the innovations and developments in deep time that have had lasting impact are those which enable us to change our local environment or at least limit its negative impacts and improve its provisions. A few of these, such as taming fire and the domestication of plants, represented major developments with particular beginnings in time, but the broadest of social processes lie in the constant relationship of humans with the physical and natural world, and the continuing attempt to develop new means of influencing it to our benefits.

Progress

In this book, I have been presenting narrative accounts of individual developments, or groups of developments, and sought to identify their many and diverse impacts. I have not sought to propose a more generalised theory or model. The arguments advanced can be tested for how well they fit into the more general historical narratives of other authors. Such ambitious broad-sweep surveys tend, overall, to be accurate in the evidence they cite from archaeology, history and related subjects, although inevitably they must be selective. They then weave these into their particular perspective and interpretation of the broad sweep of the human story. Some of the more accessible such surveys include those by the American biologist Jared Diamond, the Israeli historian Yuval Noah Harari, the British archaeologist David Wengrow with American anthropologist David Graeber, American archaeologist Robert L. Kelly, with British/Australian prehistorian Peter Bellwood and British historian Peter Frankopan making new additions to the genre.[5]

Such surveys, and those which are confined to the history of literate civilisations, formally or by implication involve the concept of progress. As a term, 'progress' has so wide a range of uses as to render challenging any assessment of its application to historical development. In Western Christian society, there was perhaps once a greater confidence in the meaning of progress. A Google Ngram test shows a gradual decline in the use of the term in English printed sources since the high point of 1850, with a continuing notable reduction year by year since 1960. Nevertheless, in political discourse, social debate and conversation, there is a general expectation that progress is a recognisable thing representing desirable and achievable targets.

If progress in human history is the betterment of the human experience, does innovation necessarily represent progress? The introduction to this book referred to the overused phrase about history 'continuity and change'. The chapters which followed describe innovations that brought about major changes in human societies, leading to long-term continuities in their use. How much are we obsessed with the idea of innovation?

Archaeology makes continuity the basis of usable classifications of a culture, with change or contrast the marker of a new cultural unit in space or time. Considering the subjectivity of these norms, prehistorian Catherine J. Frieman has emphasised how culturally determined is our modern enthusiasm for innovation.[6] There is an argument that archaeologists' analyses focus too much on success; there is need also for an archaeology of failure.[7]

Only perhaps since the beginnings of the industrial revolution in 18th-century Europe have priority and praise been given to the concept of innovation as such. In other times and places, a mood of change has been seen as a threat. With the growth of international maritime trade in the early modern period and later colonial expansion, innovation was not universally welcomed by existing rulers outside of Europe. Tokugawa Japan had already chosen isolation from the 17th century. Late Ming and early Qing emperors of China from the 15th century sought to protect their lands from European influence with an isolationist policy. In face of the reality of colonisation by European powers, local elites might be divided between those who sided with the colonisers in their claim to be advancing progress with their innovations and those who sought to defend their traditional patterns. Many ancient (or not so ancient) dynasties asserted their right to rule on the basis of long-term continuity, not innovation, change and progress.

In many countries of Europe itself before the industrial revolution, innovation had typically been seen as subversive, undermining the powers of civil rule or religious leadership. The values placed on 'tradition' can appear more powerful in the long trajectory of human history than the values placed on 'innovation'. Political authority today vacillates between 'conservative' and 'progressive' parties in many democratic industrialised states.

We might therefore distinguish between innovation as a process of change and progress as a measure of welfare or movement towards an agreed goal. When we look back at history, or even further into deep time, we need to be clear what is a contributor to progress and what is merely change. Nevertheless, common if unsaid is an assumption is that our own civilisation or society must be the pinnacle of human achievement and that elements of the past can be measured by the stages of progress towards the features we recognise as strengths today.

This reflects the largely secular assumptions of individuals in societies described by the acronym WEIRD: 'Western, educated, industrialised, rich and democratic', which would include that of the present author. This takes the present world, though not perfect, as the best in world history so far. It commonly accepts an historical sequence from the first civilisations of the ancient Middle East, through the Classical world of Greece and Rome, perhaps a brief nod to the role of the Islamic world in maintaining Classical learning, to the development of medieval into early modern states of Christian Europe followed by the expansion of European culture through the colonial enterprise into the Americas, Africa, the Pacific and parts of Asia. There are many assumptions to reinforce this image of world history such as today's dominance of the English language in communications and the spread of European clothing styles. Is this a useful working model of the past? Is it unnecessarily Eurocentric? Is it racist?

The model conceded a cyclical pattern with dips in the flow of civilisation: 'intermediate periods' in pharaonic Egypt, dark ages in Europe, conquests by Mongols and Huns – but with the assumption that each new period of civilisation progressed to a higher level than its predecessor. Such traditional world history describes a rise then fall of civilisations and their empires and cultures, whether Inca or Aztec, Babylonian or Egyptian, Greek or Persian, Roman or Mughal. Suggesting our own Western European-derived civilisation will inevitably fall is often dismissed as eccentric. When American political scientist Francis Fukuyama suggested American constitutional democracy combined with international military strength marked 'the end of history', there was both agreement and dissent; before long, the author changed his own analyses, writing about political decay in Western democracy.[8]

The rise of China, soon to become the largest economy in the world without a change in political structure, is a reminder of alternate models of human history and progress. A Chinese narrative might begin with the Shang dynasty of the 2nd millennium BC, lead through the Qin unification of China in 221 BC and the continuity of rule over a vast area despite dips in the centralised control of emperors. Progress would then be defined with the modernising republican movement of the early 20th century and continuity of the Communist Party revolution bringing the sequence to the present day.

The last century saw other definitions of human progress – racial (like those of Nazi Germany) or nationalist (like the rise of a new Russian historiography in the post-communist era).

Western scholars in history, and even more in archaeology, have been putting greater emphasis on the rights and perspectives of Indigenous people in relation to their own history. This has developed into a postmodern declaration that the historical perspectives held by such communities may have as much value as those emerging from scientific and scholarly research. But what may be morally commendable could be intellectually dubious. In reality, archaeologists may find themselves downplaying the role of their own science when engaging with the local perspectives of communities with whom they work, such as Indigenous peoples of North America or Australia, who might present a quite different story of historical development. But this tolerance for alternative views dissipates when scholars confront popular televised series endorsing ancient aliens, lost continents and imaginative chronologies to attract a wide popular audience. I have discussed elsewhere the quandary between suggesting nobly there are multiple alternate ways to describe history when acknowledging Indigenous perspectives and being distressed by the success of what we conventionally call pseudoarchaeology and pseudohistory.[9]

Arguments from archaeology and physical anthropology (with comparisons from social anthropology) have suggested that even that major change in human prehistory, from a foraging economy to settled village agriculture had negative impacts that challenge the idea that farming marked progress. In many contexts, a diet of cultivated foods produced negative impacts on human health; closer living in farming communities encouraged communicable diseases; and the time which foraging societies have for leisure rather than economic activity was substantially reduced by the commitments of agricultural life. The shift from village agriculture to urban civilisation had other impacts which may not be universally considered as progress, notably the new disparities of wealth and power. Today, we describe the industrial revolution with a mixture of admiration for the technological achievements and dismay at many of its early social impacts.

One dominant current discourse might appear to challenge the idea of our modern era representing the crowning point of human progress. Concern about the impact of human-generated carbon emissions on world climate have joined other environmental issues. These include anxiety about the speed and impact of increasing world populations, questions about genetically modified foods, reappraisal of chemical treatments in farming of plants and animals, concern over the ecological results from the replacement of native forests by plantations and so on. Such discussions can be a challenge to the tradition which saw human mastery of the natural environment as a measure of progress.

Some common measures of progress might appear to have universal value but can still encounter negative and critical responses. Current wars are smaller and more localised, if often longer lasting, than the major conflicts of the 20th century, but the ever-increasing power and sophistication of military equipment might make those in peace movements question whether this leaves us in a better position.

Undoubtedly, the contemporary level of scientific knowledge, medical expertise, community health, life expectancy, access to education and the widespread if not universal erosion of discrimination on the basis of gender, biological or ethnic classification, sexual orientation or gender identity, would seem to many of us to mark progress. But these are not universally acknowledged as progress, and citizens with a commitment to what they see as traditional religious values and obligations raise questions.

Conservative Christian groups in the United States actively challenge some changes to the legal and social norms. Those we describe as Islamic fundamentalists can cite their perception of religious requirements to resist what they see as modes imposed by external and hostile values. There are groups who have turned their back on medical science, refusing vaccinations or blood transfusions. Some in economics and administration might see a challenge if not a danger as the expense and success of medical treatment continues to extend the years of life and the costs of maintaining those no longer in employment.

Individual as well as social assessments of measures of progress thus necessarily vary. A majority might agree that scientific knowledge, engineering ability, medical treatment and sporting ability have reached the highest level ever by the first quarter of the 21st century. More subjective would be judgements on whether the latest visual art, music, dance or literature marks the highest level of achievement in each of these fields; or whether other measures of social interaction represent the pinnacle humans could reach. One person's progress is another person's backward step. Time moves forward; to some, society does not.

Presentism

Part of the reason we may think of innovations in a positive light, and as contributions to a constant march of progress in human society, is that our focus is very much on the present day. Our natural tendency is to think of our present human society as the culmination and goal of all past developments; our achievements as the peak of human endeavour and the history of the dominant species on the planet. But assessment of what is important in human history requires us to examine and seek to modify the bias of such a 'presentist' perspective.

Our celebration of 'now' as the pinnacle of history is not universal. Religious groups, in particular, can anticipate a future point as the end of history. Jewish tradition expects the coming of a messiah; Christian beliefs, from the very beginning, have anticipated a 'Second Coming' described in varying theological terms. In secular philosophy, classic 19th-century Marxism prophesied a potential post-capitalist social order.

We have, thankfully, developed our critical awareness of racism and seek to excise it. We have brought into public gaze the contexts of sexism, in language and in life, to make that recognition transform our actions. Prejudices and preconceptions in relation to homosexual and transgender people have been addressed. Nationalism has not disappeared, but it evokes strong disapproval of its extremes. But among all our 'isms', we retain the prejudice of *presentism*: the view that the present is the most important part of the human story.

Young birds and young mammals (including young humans) are genetically predisposed to see themselves as the centre of the universe. It helps them put forward vocally their need for food and warmth. When a mother's attention moves from a toddler to a new baby, the reaction can be troubling.

We are equally likely to give our own contemporary population a special role in our perception of human development. In our own lives, our sense of the speed of change is highly subjective and foreshortened. While young adults, we are sharply aware of how our world has changed (or of how we are changing our world) in contrast to that of our grandparents. As we grow older, we note the rapidity with which our familiar environment itself is changing to that of our grandchildren. We note dramatic innovations (and disappearances) in the technologies of daily life; shifts in foodstuffs and food preferences; turmoil in the political world; population mobility. We see challenges to identity, as ethnic, religious, national and gender self-descriptions alter. It appears to us that the human world must be evolving faster than ever.

By one specific analogy, if we take the most popular form of participant history – family history and the tracing of ancestors – tracing just 7 generations back gives us 128 ancestors, at perhaps a little more than 200 years before our own birth. That reassures us as individuals of the place in society that our genetic background has provided us. But given the larger number of offspring that were typical in the 19th century, and in many parts of the world have continued through and beyond the 20th century, any one of those 128 ancestors could have descendants numbering into four figures, making us individuals far less significant.

Thinking about deep time casts things differently, and the timescale by which an archaeologist records change in prehistoric societies may be longer than the timescale used by a historian focused on literate settings.

To a sea slug, the world has two dimensions. Fortunately, as sighted mammals, humans live in a world perceived largely in three dimensions. We can explore those three dimensions in our homes, our cities and our travels through countryside and in the air. The historian sees the world in the fourth dimension of time (and to the archaeologist, that fourth dimension can be very large). Considering everything in the framework of four dimensions, and with a perspective that shrinks the significance of the present day, can be challenging and can also be humbling.

The issues of 'now' loom large: they contrast with those of previous generations and will be replaced by those of the next. Today's short-term crisis may move out of the news memory of recent floods, earthquakes, tornadoes or wildfires. Those in turn might push into the background many of the continuing wars around the world, even those initially heralded as a game changer in world politics. From the beginning of 2020 well into 2023, the COVID pandemic seemed to be transforming our world, the most severe international health and demographic crisis since the 'Spanish flu' of 1918–1920.

The arrival of COVID and responses to its spread stimulated me to explore the broader historical question of transitions. Initial reactions asserted that the impacts would be so dramatic that the world would never be the same again; that COVID would transform numerous aspects of our lives, apart from the immediate effects on individual health and social health-care systems and the collapse of numerous businesses. The power of government authorities reached unfamiliar levels, and the shift to home working and home study in many post-industrial economies marked changes expected to be long-term, bringing in permanent alterations to the world of employment and the balance between home and workplace. The confident predictions that marked the first reactions to COVID raised the question of what really does mark a transition in society, and what is seen as temporary event. Some of the initial predictions about COVID's longer-term impacts now seem excessive.

Our passionate, even necessarily preoccupied, concern with the threat of a worldwide and often fatal pandemic itself temporarily moved our attention away from an even larger and longer-term issue, that of the negative effects of human agency on climate. Here, the analysis, warnings and rhetoric imply a transition so complete that it inspired a term for a new geological age, the Anthropocene.

The proposal to label an Anthropocene era puts into sharp focus another bias alongside presentism, that of *speciesism*. To traditional religions, humankind was the centre of creation: 'God made mankind in his own

Presentism

image' to rule over all other creatures. Charles Darwin helped us abandon that image, but the idea of the centrality of humanity has returned with a vengeance with the suggestion that a whole new era of geological history be named after us.

Speciesism is a confidence that privileges our own species in nature, even in our increasingly secular society. It is an equivalent of the religious perspective that humanity has a special place in the universe, or at least in the biological world of our planet. That puts anatomically modern *Homo sapiens* at the pinnacle of the natural world: by the traditional model through an evolutionary path that led from Australopithecines to *Homo habilis*, *Homo erectus* (makers of the handaxes), other *Homo* species and eventually the modern humans who replaced Neanderthals, though interbreeding with them. We now know that our own species, which emerged after about 300,000 years ago, coexisted with a number of other hominin species, all expert stone-tool craftspeople and skilled foragers. *Homo floresiensis* thrived in South-east Asia, Denisovans occupied areas of both South-east and Inner Asia, *Homo longi* has been identified in China, with claims for other hominin species in the region, while some consider coexistence of our species with the Neanderthals in Europe may have continued later than 40,000 years ago.

It is neither necessary nor constructive to downplay the impact that will arise if the human influences on climate change are not checked, nor useful to suggest these can be completely reversed. But analysis and perhaps despair have led to the proposal, now widely cited, that we are just now on the transition point between geological periods and are entering the Anthropocene era.[10] As propaganda about the relationship between humanity and the environment, this can have value, but it could also be cited as an example of the prejudice of presentism. It might be categorised as a teleology comparable to the apocalyptic prophecies of Christian sects.

Climate change is real; human causes are real; and the urgent need for human response is real. But in strictly scientific terms, hominins have occupied the planet during eras of even greater fluctuations in temperature and rainfall during the Pleistocene. We now do have an ability to respond, not least because our technology allows us to communicate worldwide and potentially coordinate our actions.

If we consider the classification of geological eras, the position in which we place ourselves and our times become rather obvious. The established current geological time scale for our planet extends from 4,000 million years ago (mya); the Cenozoic (following the extinction of the dinosaurs) from 65 mya, within which the relatively short Quaternary era in which our hominin family developed is dated from 2.6 mya.[11] That is already conventionally divided into Pleistocene and a Holocene beginning just 12,000 years ago: a division of a very short period of the Quaternary which privileges the period in which human agriculture and settled life began. Those who identify an Anthropocene era still argue whether it began with the industrial revolution (e.g. ca. 1750), with others placing it from 1950. However, geologists in the relevant committee within the International Union of Geological Sciences cannot agree on whether the Anthropocene is a valid chrono-stratigraphic unit; in 2019, a majority said yes, in March 2024 a majority said no.[12]

The artificiality of this presentism is indicated if we convert the dates into an analogy. If the 4,000 million years of earth history were equivalent to the biblical seven days of creation, then the Cenozoic began just 2.7 hours ago, and the Quaternary began 6.5 minutes ago. The Holocene began 1.8 seconds ago, so the Anthropocene would be a quarter of a second? Is it really true that this particular planet of the solar system has been speeding up its geological stages? Or have we returned to the old religious views that make the world revolve around us. Presentism? Speciesism?

In the long sweep of the human past, continuity and tradition were the driving forces, not change. That makes the major innovations described in this book particularly significant. They each had a wide range of impacts that were numerous and widespread. They helped to permanently transform many aspects of society. Presentism emphasises the innovations of our own lifetimes; history gives us a different perception of human development and how one discovery, invention or innovation could help change our world. We will continue to debate the nature of progress and the position of our own lifetimes in the human story. I hope that this book helps to inform and inspire such discussions.

Notes

Chapter 1 Introduction

1. O. O'Neill, *A Philosopher Looks at Digital Communication*, Cambridge: Cambridge University Press, 2022, pp. 5–8.
2. K. Gotham, 'Time/Space', in P. Kivisto (ed.), *The Cambridge Handbook of Social Theory*, Cambridge: Cambridge University Press, 2020, pp. 206–226.

Chapter 2 Taming Fire

1. E. V. Komarek, 'Lightning and fire ecology in Africa', *Tall Timbers Fire Ecology Conference* 11 (1971): 473–511.
2. M. Ahrens, *Lightning Fires and Lightning Strikes*, Quincy, MA: National Fire Protection Association, 2008; S. J. Pyne, *Fire: A brief history*, Seattle: University of Washington Press, 2001, p. 6.
3. J. D. Pruetz & N. M. Herzog, 'Savanna chimpanzees at Fongoli, Senegal, navigate a fire landscape', *Current Anthropology* 58: S16 (2017): S337–350.
4. M. F. Teaford & P. S. Ungar, 'Diet and the evolution of the earliest human ancestors', *Proceedings of the National Academy of Sciences* 97 (2000): 13506–13511; A. Peterson, E. F. Abella, F. E. Grine, M. F. Teaford & P. S. Ungar, 'Microwear textures of *Australopithecus africanus* and *Paranthropus robustus* molars in relation to paleoenvironment and diet', *Journal of Human Evolution* 119 (2018): 42–63.
5. A. Sistiaga, F. Husain, D. Uribelarrea et al., 'Microbial biomarkers reveal a hydrothermally active landscape at Olduvai Gorge at the dawn of the Acheulean, 1.7 Ma', *Proceedings of the National Academy of Sciences* 117 (2020): 24720–24728.
6. L. M. Stancampiano, S. Rubio-Jara, J. Panera et al., 'Organic geochemical evidence of human-controlled fires at Acheulean site of Valdocarros II (Spain, 245 kya)', *Scientific Reports* 13, article 7119 (2023): 1–2.
7. S. R. James, R. W. Dennell, A. S. Gilbert et al., 'Hominid use of fire in the Lower and Middle Pleistocene: A review of the evidence', *Current Anthropology* 30 (1989): 1–26; W. Roebroeks & P. Villa, 'On the earliest evidence for habitual use of fire in Europe', *Proceedings of the National Academy of Sciences* 108 (2011): 5209–5214.

8. S. Hlubik, R. Cutts, D. R. Braun et al., 'Hominin fire use in the Okote member at Koobi Fora, Kenya: New evidence for the old debate', *Journal of Human Evolution* 133 (2019): 214–229; Z. Stepka, I. Azuri, L. K. Horwitz, M. Chazan & F. Natalio, 'Hidden signatures of early fire at Evron Quarry (1.0 to 0.8 Mya)', *Proceedings of the National Academy of Sciences* 119 (2022): e2123439119.
9. N. Goren-Inbar, N. Alperson, M. E. Kislev et al., 'Evidence of hominin control of fire at Gesher Benot Ya'aqov, Israel', *Science* 304 (2004): 725–727; N. Alperson-Afil & N. Goren-Inbar, *The Acheulian Site of Gesher Benot Ya'aqov Volume II: Ancient flames and controlled use of fire*, Dordrecht: Springer, 2010; I. Zohar, N. Alperson-Afil, N. Goren-Inbar et al., 'Evidence for the cooking of fish 780,000 years ago at Gesher Benot Ya'aqov, Israel', *Nature Ecology & Evolution* 6 (2022): 2016–2028.
10. K. MacDonald, F. Scherjon, E. van Keen, K. Vaesen & W. Roebroeks, 'Middle Pleistocene fire use: The first signal of widespread cultural diffusion in human evolution', *Proceedings of the National Academy of Sciences* 118 (2021): e2101108118.
11. R. Wrangham, *Catching Fire: How cooking made us human*, New York: Basic Books 2009, pp. 102–103, 193–194.
12. J. Rosell & R. Blasco, 'The early use of fire among Neanderthals from a zooarchaeological perspective', *Quaternary Science Reviews* 217 (2019): 268–283.
13. R. Barkai, J. Rosell, R. Blasco & A. Gopher, 'Fire for a reason: Barbecue at Middle Pleistocene Qesem Cave, Israel', *Current Anthropology* 58: S16 (2017): S314–328; C. Lemorini, E. Cristiani, S. Cesaro et al., 'The use of ash at Late Lower Paleolithic Qesem Cave, Israel: An integrated study of use-wear and residue analysis', *PLOS ONE* 15 (2020): e0237502.
14. H. L. Dibble, D. Sandgathe, P. Goldberg, S. McPherron & V. Aldeias, 'Were Western European Neandertals able to make fire?', *Journal of Paleolithic Archaeology* 1 (2018): 54–79.
15. A. C. Sorensen, E. Claud, & M. Soressi. 'Neandertal fire-making technology inferred from microwear analysis', *Scientific Reports* 8 (2018): 10065; M. Lombard & P. Gärdenfors, 'Minds on fire: Cognitive aspects of early firemaking and the possible inventors of firemaking kits', *Cambridge Archaeological Journal* 33 (2023): 499–519.
16. J. M. M. J. G. Aarts, G. M. Alink, H. J. Franssen & W. Roebroeks, 'Evolution of hominin detoxification: Neanderthal and modern human Ah receptor respond similarly to TCDD', *Molecular Biology and Evolution* (2020): msaa287; A. G. Henry, 'Neanderthal cooking and the costs of fire', *Current Anthropology* 58: S16 (2017): S32–336.
17. H. L. Dibble, A. Abodolahzadeh, V. Aldeias et al., 'How did hominins adapt to Ice Age Europe without fire?', *Current Anthropology* 58: S16 (2017): S278–287.
18. D. Richter, R. Grün, J.-B. Renaud et al., 'The age of the hominin fossils from Jebel Irhoud, Morocco, and the origins of the Middle Stone Age', *Nature* 546 (2017): 293–296; J.-J. Hublin, A. Ben-Ncer, S. E. Bailey et al., 'New fossils from Jebel Irhoud, Morocco and the pan-African origin of *Homo sapiens*', *Nature* 546 (2017): 289–292.

19. I. Théry-Parisot, 'Gathering of firewood during the Palaeolithic', in S. Thiébault (ed.), *Charcoal Analysis: Methodological approaches, palaeoecological results and wood uses*, Oxford: Archaeopress, 2002: 243–249; C. Byrne, E. Dotte-Sarout & V. Winton, 'Charcoals as indicators of ancient tree and fuel strategies: An application of anthracology in the Australian Midwest', *Australian Archaeology* 77 (2013): 94–106.
20. Among over thirty Middle Stone Age sites in Southern Africa with evidence of fire, some hearths at Sibudu in South Africa were shown to have repeated use, others to have been used just a single time: S. E. Bentsen, 'Size matters: Preliminary results from an experimental approach to interpret Middle Stone Age hearths', *Quaternary International* 270 (2012): 95–102; S. E. Bentsen, 'Using pyrotechnology: Fire-related features and activities with a focus on the African Middle Stone Age', *Journal of Archaeological Research* 22 (2014): 141–175. European sites of the Aurignacian Upper Palaeolithic in western France provide illustration of hearths in excavated fire pits dating about 38–36,000 years ago. Calibrated dates: R. White, R. Mensan, A. E. Clark et al., 'Technologies for the control of heat and light in the Vézère Valley Aurignacian', *Current Anthropology* 58: S16 (2017): S288–302. Ninety Aurignacian hearths in Klisoura cave in the Greek Peloponnese included some constructed as clay-lined basins, of which a number also had a formed clay rim. Ash, burnt bones, plant remains and seeds were found with the hearths, and the repeated use of the same hearth site was likely: P. Karkanas, M. Koumouzelis, J. K. Kozlowski et al., 'The earliest evidence for clay hearths: Aurignacian features in Klisoura Cave 1, southern Greece', *Antiquity* 78 (2004): 513–525.
21. Y. Kedar, G. Kedar & R. Barkai, 'The influence of smoke density on hearth location and activity areas at Lower Paleolithic Lazaret Cave, France', *Scientific Reports* 12 (2022): 1–14; M. Á. Medina-Alcaide, S. Vandevelde, A. Quiles et al., '35,000 years of recurrent visits inside Nerja cave (Andalusia, Spain) based on charcoals and soot micro-layers analyses', *Scientific Reports* 13 (2023): 5901.
22. Lombard, 'Minds on fire'. A demonstration by an Australian Aboriginal man online at www.youtube.com/watch?v=JbydoLuVoZw shows the standard simple method. A cord drill is shown in use at www.youtube.com/watch?v=ZEl-Y1NvBVI. See also D. Hume, *Fire Making: The forgotten art of conjuring flame with spark, tinder, and skill*, New York: The Experiment, 2018.
23. R. B. Lee, *The !Kung San*, Cambridge: Cambridge University Press, 1979, p. 148.
24. D. Stapert & L. Johansen, 'Flint and pyrite: Making fire in the Stone Age', *Antiquity* 73 (1999): 765–777.
25. A recent study of fires at the French Upper Palaeolithic site of Abri Pataud concluded that bone was burnt accidentally, not as a fuel, while dung was certainly a fuel: F. Braadbaart, F. H. Reidsma, W. Roebroeks et al., 'Heating histories and taphonomy of ancient fireplaces: A multi-proxy case study from the Upper Palaeolithic sequence of Abri Pataud (Les Eyzies-de-Tayac, France)', *Journal of Archaeological Science: Reports* 33 (2020): 102468.

26. R. M. & C. H. Berndt, *The World of the First Australians*, 4th ed., Adelaide: Rigby, 1985, p. 118.
27. C. Turnbull, *Wayward Servants: The two worlds of the African pygmies*, London: Eyre & Spottiswoode, 1966, p. 35; J. G. Frazer, *Myths of the Origins of Fire: An essay*, London: Macmillan, 1930, pp. 42–43.
28. L. R. Binford, *Constructing Frames of Reference: An analytical method for archaeological theory building using ethnographic and environmental data sets*, Berkeley, CA: University of California Press, 2019; G. P. Murdock, 'The current status of the world's hunting and gathering peoples', in R. B. Lee & I. DeVore (eds.), *Man the Hunter*, Chicago: Aldine, 1968: 13–20.
29. S. A. De Beaune & R. White, 'Ice age lamps', *Scientific American* 268 (1993): 108–113; S. A. De Beaune, 'Palaeolithic lamps and their specialization: A hypothesis', *Current Anthropology* 28 (1987): 569–577; A. Glory, 'Le brûloir de Lascaux', *Gallia Préhistoire* 4 (1961): 174–183.
30. K. J. Gaston, T. W. Davies, S. L. Nedelec & L. A. Holt, 'Impacts of artificial light at night on biological timings', *Annual Review of Ecology, Evolution, and Systematics* 48 (2017): 49–68.
31. R. Kerkhove, *How They Fought: Indigenous tactics and weaponry of Australia's frontier wars*, Tingalpa: Boolarong Press, 2023, pp. 91–6; F. Scherjon, C. Bakels, K. MacDonald & W. Roebroeks, 'Burning the land: An ethnographic study of off-site fire use by current and historically documented foragers and implications for the interpretation of past fire practices in the landscape', *Current Anthropology* 56 (2015): 299–326.
32. Lee, *The !Kung San*, p. 108.
33. L. R. Binford, *In Pursuit of the Past: Decoding the archaeological record*, London: Thames & Hudson, 1983 (reissued Berkeley, CA: University of California Press, 2002), pp. 97, 161; N. Galanidou, 'Patterns in caves: Foragers, horticulturists, and the use of space', *Journal of Anthropological Archaeology* 19 (2000): 243–275.
34. J. B. Birdsell, 'Some predictions for the Pleistocene based on equilibrium systems among recent hunter-gatherers', in R. B. Lee & I. DeVore (eds.), *Man the Hunter*, Chicago: Aldine, 1968: 229–240.
35. Lee, *!Kung San*, p. 154.
36. Binford, *In Pursuit of the Past*, pp. 149–160.
37. T. Twomey, 'The cognitive implications of controlled fire use by early humans', *Cambridge Archaeological Journal* 23 (2013): 113–128.
38. G. C. Hillman, 'Late Palaeolithic plant foods from Wadi Kubbaniya in Upper Egypt: Dietary diversity, infant weaning, and seasonality in a riverine environment', in D. R. Harris & G. C. Hillman (eds.), *Foraging and Farming: The evolution of plant exploitation*, London: Unwin & Hyman, 1989: 207–239.
39. R. Derricourt, *Unearthing Childhood: Young lives in prehistory*, Manchester: Manchester University Press, 2018, pp. 78–85.
40. Berndt, *World of the First Australians*, p. 114.
41. Woodburn in Lee & DeVore, *Man the Hunter*, p. 51.
42. C. Lévi-Strauss, *The Raw and the Cooked*, New York: Harper & Row, 1969.
43. R. Wrangham, 'Control of fire in the Palaeolithic: Evaluating the cooking hypothesis', *Current Anthropology* 58: S16 (2017): S303–313, p. S305; R. N. Carmody,

G. S. Weintraub & R. W. Wrangham, 'Energetic consequences of thermal and nonthermal food processing', *Proceedings of the National Academy of Sciences* 48 (2011): 19199–19203. But cf. A. M. Cornélio, R. E. de Bittencourt-Navarrete, R. de Bittencourt Brum, C. M. Queiroz & M. R. Costa, 'Human brain expansion during evolution is independent of fire control and cooking', *Frontiers in Neuroscience* 10 (2016): 1–11, with tests on mice that suggested a similar calorific value from raw and cooked meat, and the implication that advanced brain function may have enabled the capture of fire, rather than the reverse.

44. Wrangham, *Catching Fire*, pp. 68–71, 77.
45. P. Goldberg, C. E. Miller, S. Schiegl et al., 'Bedding, hearths, and site maintenance in the Middle Stone Age of Sibudu Cave, KwaZulu-Natal, South Africa', *Archaeological and Anthropological Sciences* 1 (2009): 95–122; L. Wadley et al., 'Fire and grass-bedding construction 200 thousand years ago at Border Cave, South Africa', *Science* 369 (2020): 863–866.
46. J. C. Thompson, D. K. Wright, S. J. Ivory et al., 'Early human impacts and ecosystem reorganization in southern-central Africa', *Science Advances* 7 (2021): eabf9776.
47. Pyne, *Fire: A brief history*, pp. 46–64; Scherjon et al., 'Burning the land'.
48. R. Bliege Bird, D. W. Bird, B. F. Codding, C. H. Parker & J. H. Jones, 'The "fire stick farming" hypothesis: Australian Aboriginal foraging strategies, biodiversity, and anthropogenic fire mosaics', *Proceedings of the National Academy of Sciences* 105 (2008): 14796–14801, p. 14797; A. N. Williams, S. D. Mooney, S. A. Sisson & J. Marlon, 'Exploring the relationship between Aboriginal population indices and fire in Australia over the last 20,000 years', *Palaeogeography, Palaeoclimatology, Palaeoecology* 432 (2015): 49–57; see also M. Constantine IV, A. N. Williams, A. Francke et al., 'Exploration of the burning question: A long history of fire in eastern Australia with and without people', *Fire* 6, article 152 (2023).
49. G. Singh & E. A. Geissler, 'Late Cainozoic history of vegetation, fire, lake levels and climate, at Lake George, New South Wales, Australia', *Philosophical Transactions of the Royal Society B* 311 (1985): 379–447.
50. Pyne, *Fire: A brief history*, pp. 32–3; M. Hardiman, A. C. Scott, N. Pinter et al., 'Fire history on the California Channel Islands spanning human arrival in the Americas', *Philosophical Transactions of the Royal Society B* 371 (2016): 20150167.
51. Scherjon et al., 'Burning the land'.
52. R. Jones, 'Fire-stick farming', *Fire Ecology* 8 (2012): 3–8, pp. 6–7; P. Sutton & K. Walshe, *Farmers or Hunter-Gatherers? The Dark Emu debate* (Melbourne: Melbourne University Press, 2021), pp. 55–56.
53. Scherjon et al., 'Burning the land', pp. 304–305.
54. M. Domansk & J. A. Webb, 'Effect of heat treatment on siliceous rocks used in prehistoric lithic technology', *Journal of Archaeological Science* 19 (1992): 601–614.
55. K. S. Brown, C. W. Marean, A. I. R. Herries et al., 'Fire as an engineering tool of early modern humans', *Science* 325 (2009): 859–862.

56. A. Agam, I. Azuri, I. Pinkas, A. Gopher & F. Natalio, 'Estimating temperatures of heated Lower Palaeolithic flint artefacts', *Nature Human Behaviour* 5 (2021): 221–228.
57. M. Hanckel, 'Hot rocks: Heat treatment at Burrill Lake and Currarong, New South Wales', *Archaeology in Oceania* 20 (1985): 98–103; P. Schmidt & P. Hiscock, 'The antiquity of Australian silcrete heat treatment: Lake Mungo and the Willandra Lakes', *Journal of Human Evolution* 142 (2020): 102744.
58. A. R. Ennos & T. L. Chan, '"Fire hardening" spear wood does slightly harden it, but makes it much weaker and more brittle', *Biology Letters* 12 (2016): 20160174.
59. L. Wadley, T. Hodgskiss & M. Grant, 'Implications for complex cognition from the hafting of tools with compound adhesives in the Middle Stone Age, South Africa', *Proceedings of the National Academy of Sciences* 106 (2009): 9590–9594.
60. P. Pettitt, 'Darkness visible: Shadows, art and the ritual experience of caves in Upper Palaeolithic Europe', in M. Dowd & R. Hensey (eds.), *The Archaeology of Darkness*, Oxford: Oxbow, 2016: 11–23.
61. A. Pastoors & G.-C. Weniger, 'Cave art in context: Methods for the analysis of the spatial organization of cave sites', *Journal of Archaeological Research* 19 (2011): 377–400; D. Hoffmeister, 'Simulation of tallow lamp light within the 3D model of the Ardales Cave, Spain', *Quaternary International* 430 (2017): 22–29.
62. Y. Zaidner, L. Centi, M. Prévost et al., 'Middle Pleistocene *Homo* behavior and culture at 140,000 to 120,000 years ago and interactions with *Homo sapiens*', *Science* 372 (2021): 1429–1433.
63. Y. Kedar, G. Kedar & R. Barkai, 'Hypoxia in Paleolithic decorated caves: The use of artificial light in deep caves reduces oxygen concentration and induces altered states of consciousness', *Time and Mind* 14 (2021): 181–216.
64. P. B. Vandiver, O. Soffer, B. Klima & J. Svoboda, 'The origins of ceramic technology at Dolní Věstonice, Czechoslovakia', *Science* 246 (1989): 1002–1008; Z. Li et al., 'A Paleolithic bird figurine from the Lingjing site, Henan, China', *PLOS ONE* 15 (2020): e0233370.
65. D. Leder, R. Hermann, M. Hüls et al., 'A 51,000-year-old engraved bone reveals Neanderthals' capacity for symbolic behaviour', *Nature Ecology & Evolution* 5 (2021): 1273–1282.
66. A. Needham, I. Wisher, A. Langley, M. Amy & A. Little, 'Art by firelight? Using experimental and digital techniques to explore Magdalenian engraved plaquette use at Montastruc (France)', *PLOS ONE* 17 (2022): e0266146.
67. J. M. Olley, R. G. Roberts, H. Yoshida & J. M. Bowler, 'Single-grain optical dating of grave-infill associated with human burials at Lake Mungo, Australia', *Quaternary Science Reviews* 25 (2006): 2469–2474.
68. A. R. Radcliffe-Brown, *The Andaman Islanders*, Cambridge: Cambridge University Press, 1922, pp. 201–207, 258.
69. S. J. Pyne, 'Fire in the mind: Changing understandings of fire in Western civilization', *Philosophical Transactions of the Royal Society B* 317 (2016): 20150166.
70. Turnbull, *Wayward Servants*, pp. 155–156, 172, 280; M. S. Mosko, 'The symbols of "forest": A structural analysis of Mbuti culture and social organization', *American Anthropologist* 89 (1987): 896–913, pp. 901–903.
71. Frazer, *Myths of the Origins of Fire*, p. 13.

Chapter 3 Domesticating Horses

1. S. L. Olsen, 'Early horse domestication: Weighing the evidence', in S. L. Olsen, S. Grant, A. M. Choyke & L. Bartosiewicz (eds.), *Horses and Humans: The evolution of human–equine relationships*, Oxford: Archaeopress, 2006: 81–113.
2. A. O. Vershinina, P. D. Heintzman, D. G. Froese et al., 'Ancient horse genomes reveal the timing and extent of dispersals across the Bering Land Bridge', *Molecular Ecology* 30 (2021): 6144–61.
3. S. D. Webb & A. C. Hemmings, 'Last horses and first humans in North America', in Olsen et al. (eds.), *Horses and Humans*: 11–23.
4. M. Pope, S. Parfitt & M. Roberts, *The Horse Butchery Site: A high resolution record of Lower Palaeolithic hominin behaviour at Boxgrove, UK*, Woking: Spoilheap Publications, 2020; N. J. Conard, J. Serangeli, U. Böhner et al., 'Excavations at Schöningen and paradigm shifts in human evolution', *Journal of Human Evolution* 89 (2015): 1–17.
5. W. Müller, 'One horse or a hundred hares? Small game exploitation in an Upper Palaeolithic context', in J.-P. Brugal & J. Desse (eds.), *Petits animaux et sociétés humaines du complément alimentaire aux ressources utilitaires*, Antibes: Editions APDCA, 2004: 231–240.
6. J. R. Stewart, 'Neanderthal–modern human competition? A comparison between the mammals associated with Middle and Upper Palaeolithic industries in Europe during OIS 3', *International Journal of Osteoarchaeology* 14 (2004): 178–189; D. West, 'Horse hunting in central Europe at the end of the Pleistocene', in Olsen et al. (eds.), *Horses and Humans*: 25–47.
7. S. L. Olsen, 'Solutré: A theoretical approach to the reconstruction of Upper Palaeolithic hunting strategies', *Journal of Human Evolution* 18 (1989): 295–327.
8. R. D. Guthrie, 'Human–horse relations using Paleolithic art', in Olsen et al. (eds.), *Horses and Humans*: 61–77.
9. M. Pruvost, R. Bellone, N. Benecke et al., 'Genotypes of predomestic horses match phenotypes painted in Paleolithic works of cave art', *Proceedings of the National Academy of Sciences* 108 (2011): 18626–18630, p. 18626; G. Sauvet & A. Wlodarczyk, 'Eléments d'une grammaire formelle de l'art pariétal paléolithique', *L'Anthropologie* 99 (1995): 193–211, p. 195.
10. R. Pigeaud, 'Determining style in Palaeolithic cave art: A new method derived from horse images', *Antiquity* 81 (2007): 409–422.
11. Pruvost et al., 'Genotypes of predomestic horses'.
12. Apart perhaps from Spain and central Europe. P. Kelekna, *The Horse in Human History*, New York: Cambridge University Press, 2009, pp. 16–17.
13. A. Fages, K. Hanghøj, N. Khan et al., 'Tracking five millennia of horse management with extensive ancient genome time series', *Cell* 177 (2019): 1419–1435; L. Orlando, 'Ancient genomes reveal unexpected horse domestication and management dynamics', *BioEssays* 42 (2020): 1900164.
14. M. Cieslak, M. Pruvost, N. Benecke et al., 'Origin and history of mitochondrial DNA lineages in domestic horses', *PLOS ONE* 5 (2010): e15311. A more isolated wild population apparently bred in Iberia, away from that of the Eurasian steppe area where the horse was first domesticated.

15. V. Warmuth, A. Eriksson, M. A. Bower et al., 'Reconstructing the origin and spread of horse domestication in the Eurasian steppe', *Proceedings of the National Academy of Sciences* 109 (2012): 8202–8206.
16. A. N. Lau, L. Peng, H. Goto et al., 'Horse domestication and conservation genetics of Przewalski's horse inferred from sex chromosomal and autosomal sequences', *Molecular Biology and Evolution* 26 (2009): 199–208.
17. Orlando, 'Ancient genomes'.
18. A. Ludwig, M. Pruvost, M. Reissmann et al., 'Coat color variation at the beginning of horse domestication', *Science* 324 (2009): 485–485.
19. C. Gaunitz, A. Fages, K. Hanghøj et al., 'Ancient genomes revisit the ancestry of domestic and Przewalski's horses', *Science* 360 (2018): 111–114.
20. P. Librado, N. Khan, A. Fages et al., 'The origins and spread of domestic horses from the Western Eurasian steppes', *Nature* 598 (2021): 634–640.
21. Cieslak et al., 'Origin and history'; J. Lira, A. Linderholm, C. Olaria et al., 'Ancient DNA reveals traces of Iberian Neolithic and Bronze Age lineages in modern Iberian horses', *Molecular Ecology* 19 (2010): 64–78; Orlando, 'Ancient genomes'; Fages et al., 'Tracking five millennia'. Domestic horses of the agricultural Copper Age in Iberia have strong genetic links back to wild horses within that region, while genetic studies of domesticated horses in prehistoric Hungary and the Iberian peninsula supported the image of traded tame horses breeding with a strong local wild horse population, or even an independent Iberian process of domestication.
22. G. M. Matuzeviciute, E. Lightfoot, X. Liu et al., 'Archaeobotanical investigations at the earliest horse herder site of Botai in Kazakhstan', *Archaeological and Anthropological Sciences* 11 (2019): 6243–6258.
23. Olsen, 'Horse domestication', p. 262.
24. S. Olsen, B. Bradley, D. Maki & A. Outram, 'Community organisation among Copper Age sedentary horse pastoralists of Kazakhstan', in D. L. Peterson, L. M. Popova & A. T. Smith (eds.), *Beyond the Steppe and the Sown*, Leiden: Brill 2006: 89–111.
25. A. K. Outram, N. A. Stear, R. Bendrey et al., 'The earliest horse harnessing and milking', *Science* 323 (2009): 1332–1335; D. Brown & D. Anthony, 'Bit wear, horseback riding and the Botai site in Kazakstan', *Journal of Archaeological Science* 25 (1998): 331–347; W. T. T. Taylor & C. I. Barrón-Ortiz, 'Rethinking the evidence for early horse domestication at Botai', *Scientific Reports* 11 (2021): 7440.
26. C. French & M. Kousoulakou, 'Geomorphological and micromorphological investigations of palaeosols, valley sediments, and a sunken-floored dwelling at Botai, Kazakhstan', in M. Levine, C. Renfrew & K. Boyle (eds.), *Prehistoric Steppe Adaptation and the Horse*, Cambridge: McDonald Institute, 2003: 105–114.
27. I. V. Chechushkov & P. A. Kosintsev, 'The Botai horse practices represent the neolithization process in the central Eurasian steppes: Important findings from a new study on ancient horse DNA', *Journal of Archaeological Science: Reports* 32 (2020): 102426; A. Fages, A. Seguin-Orlando, M. Germonpré & L. Orlando, 'Horse males became over-represented in archaeological assemblages during the Bronze Age', *Journal of Archaeological Science: Reports* 31, article 102364 (2020): pp. 3–4.

28. S. Olsen, 'The exploitation of horses at Botai, Kazakhstan', in Levine, Renfrew & Boyle (eds.), *Prehistoric Steppe Adaptation*, pp. 83–103.
29. D. W. Anthony, *The Horse, the Wheel, and Language: How Bronze-Age riders from the Eurasian steppes shaped the modern world*, Princeton, NJ: Princeton University Press, 2010, p. 19.
30. S. L. Olsen, 'Early horse domestication on the Eurasian steppe', in M. A. Zeder, D. G. Bradley, E. Emshwiller & B. D. Smith (eds.), *Documenting Domestication: New genetic and archaeological paradigms*, Berkeley: University of California Press, 2006: 245–269, p. 262. Horse sacrifice and burial is also seen at Krasnyi Yar.
31. Outram et al., 'Earliest horse'.
32. M. A. Levine, 'Domestication and early history of the horse', in D. Mills & S. W. McDonnell (eds.), *The Domestic Horse: The origins, development and management of its behaviour*, Cambridge: Cambridge University Press, 2005: 5–22, pp. 15–19.
33. Gaunitz et al., 'Ancient genomes'.
34. N. Benecke & A. van den Driesch, 'Horse exploitation in the Kazakh steppes during the Eneolithic and Bronze Age', in Levine, Renfrew & Boyle (eds.), *Prehistoric Steppe Adaptation*: 69–82, p. 81.
35. Anthony, *The Horse*, p. 221; Olsen, 'Early horse domestication', p. 255; Levine, 'Domestication and early history of the horse', p. 9. In her work, Levine is sceptical of some of the claims for horse domestication, noting that arguments can be raised about interpretations of individual bit wear, cheek pieces and carved maces/sceptres.
36. Kelekna, *The Horse*; D. W. Anthony & D. R. Brown, 'Eneolithic horse exploitation in the Eurasian steppes: Diet, ritual and riding', *Antiquity* 74 (2000): 75–86; Anthony, *The Horse*, pp. 184, 201.
37. P. L. Kohl, *The Making of Bronze Age Eurasia*, Cambridge: Cambridge University Press (2007), p. 143; a ritual deposit of bones of dogs and a horse showing bit wear has now been shown to be intrusive and late.
38. D. Y. Telegin, *Dereivka: A settlement and cemetery of Copper Age horse keepers on the Middle Dnieper*, Oxford: BAR, 1986; R. Drews, *Early Riders: The beginnings of mounted warfare in Asia and Europe*, London: Routledge, 2004, pp. 13–14; Levine, 'Domestication and early history of the horse', p. 9.
39. S. Mileto, E. Kaiser, Y. Rassamakin & R. P. Evershed, 'New insights into the subsistence economy of the Eneolithic Dereivka culture of the Ukrainian North-Pontic region through lipid residues analysis of pottery vessels', *Journal of Archaeological Science: Reports* 13 (2017): 67–74; S. Mileto, E. Kaiser, Y. Rassamakin, H. Whelton & R. P. Evershed, 'Differing modes of animal exploitation in North-Pontic Eneolithic and Bronze Age societies', *STAR: Science & Technology of Archaeological Research* 3 (2017): 112–125.
40. M. A. Levine, 'Botai and the origins of horse domestication', *Journal of Anthropological Archaeology* 18 (1999): 29–78; Anthony, *The Horse*, p. 205.
41. Drews, *Early Riders*, p. 25; S. Wilkin, A. Ventresca Miller, R. Fernandes et al., 'Dairying enabled Early Bronze Age Yamnaya steppe expansions', *Nature* 598 (2021): 629–633.
42. M. Trautmann et al., 'First bioanthropological evidence for Yamnaya horsemanship', *Science Advances* 9 (2023): eade2451.

43. P. de Barros Damgaard et al., 'The first horse herders and the impact of early Bronze Age steppe expansions into Asia', *Science* 360 (2018): eaar7711.
44. Librado et al., 'Origins and spread'.
45. Anthony, *The Horse*, pp. 203–204. Decrease in the size of horses and increase in variability after ca. 2500 BC suggest they were tame.
46. J. A. Wilson, R. S. Weiner, J. P. Carzoli & R. Kram, 'Were timbers transported to Chaco using tumplines? A feasibility study', *Journal of Archaeological Science: Reports* 48 (2023): 103876.
47. S. Piggott, *Wagon, Chariot and Carriage: Symbol and status in the history of transport*, London: Thames & Hudson, 1992, p. 16.
48. J. Bakker, J. Kruk, A. E. Lanting & S. Milisauskas, 'The earliest evidence of wheeled vehicles in Europe and the Near East', *Antiquity* 73 (1999): 778–790; S. Burmeister, 'Early wagons in Eurasia: disentangling an enigmatic innovation', in P. W. Stockhammer & J. Maran (eds.), *Appropriating Innovations: Entangled knowledge in Eurasia, 5000–1500 BCE*, Oxford: Oxbow, 2017: 69–77; J. H. Crouwel, 'Wheeled vehicles and their draught animals in the Ancient Near East: An update', in P. Raulwing, K. M. Linduff & J. H. Crouwel (eds.), *Equids and Wheeled Vehicles in the Ancient World*, Oxford: BAR, 2019: 29–48.
49. A. Sherratt, 'Plough and pastoralism: Aspects of the secondary products revolution', in I. Hodder, G. Isaac & N. Hammond (eds.), *Pattern of the Past*, Cambridge: Cambridge University Press, 1981: 261–306.
50. Piggott, *Wagon, Chariot and Carriage*, p. 17.
51. Burmeister, 'Early wagons', p. 71.
52. Piggott, *Wagon, Chariot and Carriage*, p. 14.
53. M. A. Littauer and J. H. Crouwel, *Wheeled Vehicles and Ridden Animals in the Ancient Near East*, Leiden: Brill, 1979, pp. 13ff.
54. Littauer & Crouwel, *Wheeled Vehicles*, pp. 15–16, 35, 44.
55. J. Maran, 'Wheels of change: the polysemous nature of early wheeled vehicles in 3rd millennium BCE central and northwest European society', in Stockhammer and Maran (eds.), *Appropriating Innovation*: 109–121.
56. Anthony, *The Horse*, pp. 311–313.
57. Kohl, *The Making*, p. 119.
58. Piggott, *Wagon, Chariot and Carriage*, pp. 21–23; and, he noted, almost identical to the gauge of British Railways.
59. Anthony, *The Horse*, p. 91. A Proto-Indo-European (PIE) language may have existed in the late 5th millennium in a finite area, and that formative language had ceased by 2500 BC.
60. L. B. Kirtcho, 'The earliest wheeled transport in Southwestern Central Asia: New finds from Altyn-Depe', *Archaeology, Ethnology and Anthropology of Eurasia* 37 (2009): 25–33.
61. J. H. Crouwel, 'Metal wheel tyres from the ancient Near East and Central Asia', *Iraq* 74 (2012): 89–95.
62. E. T. Todd, L. Tonasso-Calvière, L. Chauvey et al., 'The genomic history and global expansion of domestic donkeys', *Science* 377 (2022): 1172–1180. See also S. Rossel, F. Marshall, J. Peters et al., 'Domestication of the donkey: Timing, processes, and indicators', *Proceedings of the National Academy of Sciences* 105

(2008): 3715–3720; and P. Mitchell, *The Donkey in Human History: An archaeological perspective*, Oxford: Oxford University Press, 2018.
63. L. K. Horwitz, E. Tchernov, P. Ducos et al., 'Animal domestication in the southern Levant', *Paléorient* 25 (1999): 63–80, suggests wild donkeys may be present in bone deposits from the 8th–7th millennia BC in Jordan. However, Horwitz is one of the authors of the 2022 genome study by Todd et al. which argues for a later spread of the domesticated donkey.
64. E. A. Bennett, J. Weber, W. Bendhafer et al., 'The genetic identity of the earliest human-made hybrid animals, the kungas of Syro-Mesopotamia', *Science Advances* 8 (2022): eabm0218.
65. K. Jones-Bley, 'The evolution of the chariot', in Olsen et al. (eds.), *Horses and Humans*: 181–192. On the development of harnessing, see G. Brownrigg, 'Harnessing the chariot horse', in Raulwing, Linduff & Crouwel (eds.), *Equids and Wheeled Vehicles*: 85–96.
66. Piggott, *Wagon, Chariot and Carriage*, p. 42ff.
67. G. Brownrigg, 'Horse control and the bit', in Olsen et al. (eds.), *Horses and Humans*: 165–171.
68. P. Raulwing, *Horses, Chariots and Indo-Europeans*, Budapest: Archaeolingua, 2000 invokes language analysis to argue for an initial development in the state societies of the Near East.
69. P. F. Kuznetsov, 'The emergence of Bronze Age chariots in eastern Europe', *Antiquity* 80 (2006): 638–645; I. V. Chechushkov & A. V. Epimakhov, 'Eurasian steppe chariots and social complexity during the Bronze Age', *Journal of World Prehistory* 31 (2018): 435–483; A. Sherratt, 'The horse and the wheel: The dialectics of change in the circum-Pontic region and adjacent areas, 4500–1500 BC', in Levine, Renfrew & Boyle (eds.), *Prehistoric Steppe Adaptation*: 233–252; E. A. Cherlenok, 'The chariot in Bronze Age funerary rites of the Eurasian steppes', in Olsen et al. (eds.), *Horses and Humans*: 173–179, p. 177.
70. Chechushkov & Epimakhov, 'Eurasian steppe chariots'; E. Kupriyanova, A. Epimakhov, N. Berseneva & A. Bersenev, 'Bronze Age charioteers of the Eurasian Steppe: A part-time occupation for select men?', *Praehistorische Zeitschrift* 92 (2017): 40–65.
71. S. Lindner, 'Chariots in the Eurasian Steppe: A Bayesian approach to the emergence of horse-drawn transport in the early second millennium BC', *Antiquity* 94 (2020): 361–380, p. 377.
72. Kelekna, *The Horse*, p. 51.
73. E. E. Kuzmina, 'Origins of pastoralism in the Eurasian steppe', in Levine, Renfrew & Boyle (eds.), *Prehistoric Steppe Adaptation*: 203–232.
74. Benecke & van den Driesch, 'Horse exploitation'.
75. Chechushkov & Epimakhov, 'Eurasian steppe chariots', p. 441. By the late 2nd millennium, the rod-like cheek piece was in use widely from the Carpathians as far east as the Altai.
76. Fages et al., 'Horse males'.
77. Rock carvings show the chariot in Scandinavia and north Italy by the 12th century BC; in the Hallstatt C Iron Age of central Europe, the chariot was found

in grave offerings in the 7th century BC. Burials of horses and chariots were present in Italy by the 1st millennium.
78. E. T. Shev, 'The introduction of the domesticated horse in Southwest Asia', *Archaeology, Ethnology & Anthropology of Eurasia* 44 (2016): 123–136.
79. J. Oates, 'A note on the early evidence for horse in western Asia', in Levine, Renfrew & Boyle (eds.), *Prehistoric Steppe Adaptation*: 115–125, pp. 117–119.
80. P. R. S. Moorey, 'The emergence of the light, horse-drawn chariot in the Near-East c. 2000–1500 BC', *World Archaeology* 18 (1986): 196–215.
81. Moorey, 'The emergence', pp. 201–202.
82. Kelekna, *The Horse*, pp. 95–97.
83. M. S. Drower, 'The domestication of the horse', in P. Ucko & G. W. Dimbleby (eds.), *The Domestication and Exploitation of Plants and Animals*, London: Routledge, 2017: 471–478, p. 473.
84. Kelekna, *The Horse*, p. 300.
85. Piggott, *Wagon, Chariot and Carriage*, p. 57.
86. Chechushkov & Epimakhov, 'Eurasian steppe chariots', p. 444.
87. Olsen, 'Early horse domestication', p. 249.
88. J. H. Crouwel, *Chariots and Other Wheeled Vehicles in Italy before the Roman Empire*, Oxford: Oxbow, 2012, p. 52.
89. C. Zhang, Y. Wang, J. Zhang et al., 'Elite chariots and early horse transport at the Bronze Age burial site of Shijia', *Antiquity* 97 (2023): 636–653.
90. T. G. E. Powell, 'The introduction of horse-riding to temperate Europe: A contributory note', *Proceedings of the Prehistoric Society* 37 (1971): 1–14; Oates, 'A note'.
91. Oates, 'A note'; Littauer & Crouwel, *Wheeled Vehicles*, pp. 45–46.
92. P. R. S. Moorey, 'Pictorial evidence for the history of horse-riding in Iraq before the Kassite period', *Iraq* 32 (1970): 36–50.
93. Piggott, *Wagon, Chariot and Carriage*, pp. 69ff.
94. R. N. Spengler, 'Agriculture in the Central Asian Bronze Age', *Journal of World Prehistory* 28 (2015): 215–253.
95. S. Wilkin, A. Ventresca Miller, W. T. T. Taylor et al., 'Dairy pastoralism sustained eastern Eurasian steppe populations for 5,000 years', *Nature Ecology & Evolution* 4 (2020): 346–355; W. T. T. Taylor, J. Clark, J. Bayarsaikhan et al., 'Early pastoral economies and herding transitions in Eastern Eurasia', *Scientific Reports* 10 (2020): 1001.
96. W. Taylor, J. Cao, W. Fan et al., 'Understanding early horse transport in eastern Eurasia through analysis of equine dentition', *Antiquity* 95 (2021): 1478–1494, p. 1480.
97. Taylor et al., 'Understanding', p. 1491.
98. Piggott, *Wagon, Chariot and Carriage*, p. 65; U. L. Dietz, 'Horseback riding: Man's access to speed?', in Levine, Renfrew & Boyle (eds.), *Prehistoric Steppe Adaptation*: 189–201.
99. Kelekna, *The Horse*.
100. J. Rawson, L. Huan & W. M. T. Taylor, 'Seeking horses: Allies, clients and exchanges in the Zhou Period (1045–221 BC)', *Journal of World Prehistory* 34 (2021): 489–530.

101. V. Mair, 'The horse in late prehistoric China: Wresting culture and control from the "barbarians"', in Levine, Renfrew & Boyle (eds.), *Prehistoric Steppe Adaptation*: 163–188; K. M. Linduff, 'A walk on the wild side: Late Shang appropriation of horses in China', in Levine, Renfrew & Boyle (eds.), *Prehistoric Steppe Adaptation*: 139–162.
102. A. Curry, 'How the horse powered human prehistory', *Science* 370 (2020): 646–647; Y. Li, C. Zhang, W. T. T. Taylor et al., 'Early evidence for mounted horseback riding in northwest China', *Proceedings of the National Academy of Sciences* 117 (2020): 29569–29576.
103. Piggott, *Wagon, Chariot and Carriage*, p. 63.
104. S. Hu, Y. Hu, J. Yang et al., 'From pack animals to polo: Donkeys from the ninth-century Tang tomb of an elite lady in Xi'an, China', *Antiquity* 94 (2020): 455–472.
105. Genetic evidence suggests substantial population admixture west and east within groups classified as Xiongnu after about 200 BC: Curry, 'How the horse'.
106. M. A. Levine, 'Eating horses: The evolutionary significance of hippophagy', *Antiquity* 72 (1998): 90–100.
107. M. Levine, 'The origin of horse husbandry on the Eurasian steppe', in M. Levine, Y. Rassamakin, A. Kislenko & N. Tatarintseva (eds.), *Late Prehistoric Exploitation of the Eurasian Steppe*, Cambridge: McDonald Institute, 1999: 5–58.
108. S. Leteux, 'Is hippophagy a taboo in constant evolution?', *Menu: Journal of Food and Hospitality*, 1 (2012): 1–13.
109. https://en.wikipedia.org/wiki/Horse_meat, citing Food and Agriculture Organisation statistics.
110. Olsen, 'Early horse domestication', p. 264.
111. Levine, 'Eating horses', p. 93.
112. V. B. Kovalevskaya, 'Turning points in horse breeding in the Eurasian Steppes and the Near East', *Archaeology, Ethnology and Anthropology of Eurasia* 47 (2019): 33–41.
113. P. Mitchell, *Horse Nations: The worldwide impact of the horse on Indigenous societies post-1492*, Oxford: Oxford University Press, 2015; W. T. T. Taylor, P. Librado, M. H. T. Icu et al., 'Early dispersal of domestic horses into the Great Plains and northern Rockies', *Science* 379 (2023): 1316–1323.
114. E. West, *Contested Plains: Indians, goldseekers and the rush to Colorado*, Lawrence: University of Kansas Press, 1998, p. 50.
115. Kelekna, *The Horse*, p. 47.
116. Levine, 'The origin of horse husbandry'.
117. A. D. Crown, 'Tidings and instructions: How news travelled in the Ancient Near East', *Journal of the Economic and Social History of the Orient* 17 (1974): 244–271; M.-E. E. Abo-Eleaz, 'Neglect and detention of messengers in Egypt during the fourteenth and thirteenth centuries BCE', *Journal of the American Research Center in Egypt* 54 (2018): 17–34, p. 30.
118. D. W. Anthony, D. R. Brown & C. George, 'Early horseback riding and warfare', in Olsen et al. (eds.), *Horses and Humans*: 137–156, pp. 148–152.

119. P. Turchin, D. Hoyer, A. Korotayev et al., 'Rise of the war machines: Charting the evolution of military technologies from the Neolithic to the Industrial Revolution', *PLOS ONE* 16 (2021): e0258161.
120. Powell, 'The introduction'; Drews, *Early Riders*, p. 86.
121. U. L. Dietz, '"Cimmerian" bridles: Progress in cavalry technology?', in Olsen et al. (eds.), *Horses and Humans*: 157–163.
122. Powell, 'The introduction'; Crouwel, *Chariots*, p. 97.
123. Levine, 'The origin of horse husbandry'.
124. Drews, *Early Riders*, p. 91.
125. C. Renfrew, *Archaeology and Language: The puzzle of Indo-European languages*, London: Jonathan Cape, 1987, p. 139.
126. Piggott, *Wagon, Chariot and Carriage*, p. 89; A. E. Dien, 'The stirrup and its effect on Chinese military history', *Ars Orientalis* 16 (1986): 33–56; M. A. Littauer, 'Early stirrups', *Antiquity* 55 (1981): 99–105.
127. Drews, *Early Riders*; Moorey, 'The emergence', pp. 203–204.
128. L. Delpeut & C. Willekes, 'Realism as a representational strategy in depictions of horses in Ancient Greek and Egyptian Art: How purpose influences appearance', *Arts (MDPI)* 12, article 57 (2023).
129. Moorey, 'The emergence', p. 211.
130. Piggott, *Wagon, Chariot and Carriage*, p. 64.
131. H. C. Chehabi & A. Guttmann, 'From Iran to all of Asia: The origin and diffusion of polo', *International Journal of the History of Sport* 19 (2002): 384–400.
132. D. Gerhold, 'Packhorses and wheeled vehicles in England, 1550–1800', *Journal of Transport History* 14 (1993): 1–26.
133. R. W. Bulliet, *The Camel and the Wheel*, Cambridge, MA: Harvard University Press, 1975.
134. Piggott, *Wagon, Chariot and Carriage*, p. 18.
135. Crouwel, *Chariots*, p. 3.
136. J. Langdon, 'Horse hauling: A revolution in vehicle transport in twelfth-and thirteenth-century England?', *Past and Present* 103 (1984): 37–66; S. J. G. Hall, 'The horse in human society', in Mills & McDonnell (eds.), *The Domestic Horse*: 23–32, p. 25; Gerhold, 'Packhorses', pp. 3–4.
137. R. Turvey, 'Horse traction in Victorian London', *Journal of Transport History* 26 (2005): 38–59.
138. S. Piggott, 'Copper vehicle-models in the Indus civilization', *Journal of the Royal Asiatic Society of Great Britain and Ireland* 2 (1970): 200–202.
139. Piggott, *Wagon, Chariot and Carriage*, pp. 126–127, 137ff.
140. Crouwel, *Chariots*, pp. 70ff.
141. Turvey, 'Horse traction', p. 49.
142. Piggott, *Wagon, Chariot and Carriage*, p. 124.
143. Hall, 'The horse in human society', p. 25.
144. Taylor, 'Early pastoral economies', p. 12.
145. P. J. Cross, 'Horse burial in first millennium AD Britain: Issues of interpretation', *European Journal of Archaeology* 14 (2011): 190–209.

Chapter 4 Developing Writing

1. V. G. Childe, *Man Makes Himself*, London: Watts, 1936, pp. 140–178; V. G. Childe, *What Happened in History*, London: Penguin, 1942, pp. 89–102.
2. R. Fletcher, 'Urban labels and settlement trajectories', *Journal of Urban Archaeology* 1 (2020): 31–48.
3. S. D. Houston (ed.), *The First Writing: Script invention as history and process*, Cambridge: Cambridge University Press, 2004, pp. 8–10.
4. M. Liverani, *The Ancient Near East: History, society and economy*, London: Routledge, 2014: 61–92.
5. C. S. Henshilwood, F. d'Errico, K. L. van Niekerk et al., 'An abstract drawing from the 73,000-year-old levels at Blombos Cave, South Africa', *Nature* 562 (2018): 115–118.
6. R. K. Englund, 'Texts from the Late Uruk Period', in J. Bauer, R. K. Englund & M. Krebernik (eds.), *Mesopotamien: Späturuk-Zeit und Frühdynastische Zeit*, Freiburg: Universitätsverlag and Gottingen: Vandenhoeck Ruprecht, 1998: 15–236, p. 63.
7. P. M. M. G. Akkermans & K. Duistermaat, 'More seals and sealings from Neolithic Tell Sabi Abyad, Syria', *Levant* 36 (2004): 1–11.
8. S. Denham, 'The meanings of late Neolithic stamp seals in North Mesopotamia', PhD thesis, University of Manchester, 2013, pp. 40, 52, 246, 258; online at www.research.manchester.ac.uk/portal/files/54542679/full_text.pdf.
9. S. K. Costello, 'Image, memory and ritual: Re-viewing the antecedents of writing', *Cambridge Archaeological Journal* 21 (2011): 247–262.
10. D. Schmandt-Besserat, 'The envelopes that bear the first writing', *Technology and Culture* 21 (1980): 357–385; S. J. Lieberman, 'Of clay pebbles, hollow clay balls, and writing: A Sumerian view', *American Journal of Archaeology* 84 (1980): 339–358.
11. D. Schmandt-Besserat, 'Before numerals', *Visible Language* 18 (1984): 48–60. Her interpretations that early written signs developed directly from the impressions used on such clay objects have not been universally accepted.
12. M. A. Powell, 'The origin of the sexagesimal system: The interaction of language and writing', *Visible Language* 6 (1972): 5–18.
13. G. Algaze, *Ancient Mesopotamia at the Dawn of Civilization*, Chicago: University of Chicago Press, 2009, p. 137; Tablet (W19408,76): E. Robson, 'Mesopotamian mathematics', in V. Katz (ed.), *The Mathematics of Egypt, Mesopotamia, China, India, and Islam: A sourcebook*, Princeton, NJ: Princeton University Press, 2007: 57–186, pp. 73–74.
14. S. Chrisomalis, *Numerical Notation: A comparative history*, Cambridge: Cambridge University Press, 2010, pp. 34–67; M. Valério & S. Ferrara, 'Numeracy at the dawn of writing: Mesopotamia and beyond', *Historia Mathematica* 59 (2022): 35–53.
15. S. Pollock, *Ancient Mesopotamia: The Eden that never was*, Cambridge: Cambridge University Press, 1999.

16. P. Charvát, *Mesopotamia before History*, London: Routledge, 2002, pp. 98–106, 160–161; N. Crüsemann, M. van Ess, M. Hilgert & B. Salje (eds.), *Uruk: First city of the ancient world*, Los Angeles: Getty Museum, 2019. Later Uruk periods are now sometimes referred to as Later Chalcolithic 5. The chronology used here adopts that in H. Crawford (ed.), *The Sumerian World*, London: Routledge, 2012, p. xxiii.
17. Algaze, *Ancient Mesopotamia*, pp. 50–55.
18. H. J. Nissen, 'The archaic texts from Uruk', *World Archaeology* 17 (1986): 317–334; Englund, 'Texts', pp. 32, 34, 65.
19. J. S. Cooper, 'Babylonian beginnings: The origin of the cuneiform writing system in comparative perspective', in Houston (ed.), *The First Writing*: 71–99, p. 84.
20. Englund, 'Texts', pp. 66–68.
21. M. Krebernik, 'Die texte aus Fāra und Tell Abū Salābiḫ', in Bauer, Englund and Krebernik (eds.), *Mesopotamien*: 237–430.
22. J. H. Taylor, 'Tablets as artefacts, scribes as artisans', in K. Radner & E. Robson (eds.), *The Oxford Handbook of Cuneiform Culture*, Oxford: Oxford University Press, 2011, pp. 4–31, p. 13; J.-J. Glassner, *The Invention of Cuneiform: Writing in Sumer*, Baltimore, MD: Johns Hopkins University Press, 2003, pp. 118–120; Cooper, 'Babylonian beginnings', p. 85.
23. A. Seri, '"Adaptation of cuneiform to write Akkadian', in C. Woods (ed.), *Visible Language: Inventions of writing in the ancient Middle East and beyond*, Chicago: Oriental Institute of the University of Chicago, 2010: 85–93.
24. G. Leick, *Mesopotamia: The invention of the city*, London: Penguin, 2001, pp. 46–47; Charvát, *Mesopotamia before History*, p. 160.
25. E. C. Kohler, 'Prehistoric Egypt', in K. Radner, N. Moeller & D. T. Potts (eds.), *The Oxford History of the Ancient Near East: Volume I: From the beginnings to Old Kingdom Egypt and the Dynasty of Akkad*, Oxford: Oxford University Press, 2020: 95–162, pp. 128–129; J. Kahl, 'Hieroglyphic writing during the fourth millennium BC: An analysis of systems', *Archéo-Nil* 11 (2001): 101–134, pp. 103–104.
26. Kohler, 'Prehistoric Egypt', p. 106; L. Bestock, 'Early dynastic Egypt' in Radner, Moeller and Potts (eds.), *Oxford History of the Ancient Near East*: 245–315, p. 149; E. J. Macarthur, 'The conception and development of the Egyptian writing system', in Woods (ed.), *Visible Language*: 115–121, p. 116.
27. J. N. Postgate, *Early Mesopotamia: Society and economy at the dawn of history*, London: Routledge, 1992, p. 56.
28. Bestock, 'Early dynastic Egypt', p. 256; I. Regulski, 'The beginning of hieratic writing in Egypt', *Studien zur Altägyptischen Kultur* 38 (2009): 259–274, p. 259.
29. Cooper, 'Babylonian beginnings', p. 72; P. Tallet & G. Marouard, 'The harbor of Khufu on the Red Sea coast at Wadi al-Jarf, Egypt', *Near Eastern Archaeology* 77 (2014): 4–14.
30. J. D. Ray, 'The emergence of writing in Egypt', *World Archaeology* 17 (1986): 307–316, p. 308.
31. G. Dreyer, *Umm el-Qaab I: Das prädynastische Königsgrab U-j und seine frühen Schriftzeugnisse*, Mainz: Philipp van Zabern, 1998.
32. Dreyer, *Umm el-Qaab*, pp. 183–187.

33. J. Baines, 'The earliest Egyptian writing: Development, context, purpose', in Houston (ed.), *The First Writing*: 150–189, pp. 150–151, 153–154.
34. I. Regulski, 'The origins and early development of writing in Egypt', in *The Oxford Handbook of Topics in Archaeology*, https://doi.org/10.1093/oxfordhb/9780199935413.013.61 (2014), p. 8
35. Kahl, 'Hieroglyphic writing', p. 114.
36. P. Kaplony, *Die Inschriften der ägyptischen Frühzeit*, 3 vols., Wiesbaden: Otto Harrassowitz, 1963; supplemented by P. Kaplony, *Kleine Beiträge zu den Inschriften der ägyptischen Frühzeit*, Wiesbaden: Otto Harrassowitz, 1966; J. Kahl, N. Kloth & U. Zimmermann, *Die Inschriften der 3. Dynastie*, Wiesbaden: Otto Harrassowitz, 1995.
37. Regulski, 'The origins', p. 3.
38. Regulski, 'The origins', p. 16; Regulski, 'The beginning of hieratic writing'.
39. R. K. Englund, 'The state of decipherment of proto-Elamite', in Houston (ed.), *The First Writing*: 100–149, p. 140; F. Desset, 'Linear Elamite writing', in J. Álvarez-Mon, G. P. Basello & Y. Wicks (eds.), *The Elamite World*, London: Routledge, 2018: 397–415; F. Desset, K. Tabibzadeh, M. Kervran, G. P. Basello & G. Marchesi, 'The decipherment of Linear Elamite writing', *Zeitschrift für Assyriologie und vorderasiatische Archäologie* 112 (2022): 11–60.
40. For links to Egypt, see B. Haring, 'Ancient Egypt and the earliest known stages of alphabetic writing', in P. J. Boyes & P. M. Steele (eds.), *Understanding Relations between Scripts II: Early alphabets*, Oxford: Oxbow, 2020: 53–67.
41. P. Boyes, 'The social context of writing in ancient Ugarit', *The Ancient Near East Today* 10 (2022): 1–6.
42. T. A. Kohler, D. Bird & D. H. Wolpert, 'Social scale and collective computation: Does information processing limit rate of growth in scale?', *Journal of Social Computing* 3 (2022): 1–17; J. Shin, M. H. Price, D. H. Wolpert et al., 'Scale and information-processing thresholds in Holocene social evolution', *Nature Communications* 11 (2020): 1–8.
43. P. Kelly, J. Winters, H. Milton & O. Morin, 'The predictable evolution of letter shapes: An emergent script of West Africa recapitulates historical change in writing systems', *Current Anthropology* 62 (2021): 669–691.
44. R. K. Flad, 'Divination and power: A multiregional view of the development of oracle bone divination in early China', *Current Anthropology* 49 (2008): 403–437.
45. E. L. Shaughnessy, 'The beginnings of writing in China', in Woods (ed.), *Visible Language*: 215–221; R. W. Bagley, 'Anyang writing and the origin of the Chinese writing system', in Houston (ed.), *The First Writing*: 190–249.
46. P. Demattè, 'The origins of Chinese writing: The Neolithic evidence', *Cambridge Archaeological Journal* 20 (2010): 211–228.
47. X. Fu, Z. Yang, Z. Zeng, Y. Zhang & Q. Zhou, 'Improvement of oracle bone inscription recognition accuracy: A deep learning perspective', *ISPRS International Journal of Geo-Information* 11(1), article 45 (2022).
48. Shaughnessy, 'The beginnings', p. 217.
49. J. W. Palka, 'The development of Maya writing', in Woods (ed.), *Visible Language*: 225–229; J. Marcus, 'The origins of Mesoamerican writing', *Annual Review of Anthropology* 5 (1976): 35–67.

50. M. C. R. Martinez, P. O. Ceballos & M. D. Coe, 'Oldest writing in the New World', *Science* 313 (2006): 1610–1614; M. D. Carrasco & J. D. Englehardt, 'Diphrastic kennings on the Cascajal block and the emergence of Mesoamerican writing', *Cambridge Archaeological Journal* 25 (2015): 635–656; D. F. Mora-Marín, 'The Cascajal block: New line drawing, distributional analysis, and orthographic patterns', *Ancient Mesoamerica* 31 (2020): 210–229.
51. Cooper, 'Babylonian beginnings', p. 84.
52. Glassner, *The Invention*, pp. 180–183.
53. Algaze, *Ancient Mesopotamia*, pp. 137–138.
54. Englund, 'Texts', pp. 26–27.
55. T. E. Balke, 'The interplay of material, text, and iconography in some of the oldest "legal" documents', in T. E. Balke & C. Tsouparopoulou (eds.), *Materiality of Writing in Early Mesopotamia*, Berlin: De Gruyter, 2016: 73–94.
56. www.metmuseum.org/art/collection/search/329079.
57. Liverani, *Ancient Near East*, pp. 65–68.
58. Englund, 'Texts', p. 70.
59. G. Leick, *Mesopotamia*, pp. 46–47.
60. D. Graeber & D. Wengrow, *The Dawn of Everything: A new history of humanity*, London: Penguin & New York: Farrar, Straus & Giroux, 2021, pp. 288ff., 304.
61. Charvát, *Mesopotamia before History*, pp. 142, 210–211, 218–219.
62. J. Baten, G. Benati & A. Sołtysiak, 'Violence trends in the ancient Middle East between 12,000 and 400 BCE', *Nature Human Behaviour* 7 (2023): 2064–2073.
63. M. Lichtheim, *Ancient Egyptian Literature: Volume I, The Old and Middle Kingdoms*, Berkeley: University of California Press, 2006, p. 28.
64. P. Tallet, 'The Wadi el-Jarf site: A harbor of Khufu on the Red Sea', *Journal of Ancient Egyptian Interconnections* 5 (2013): 76–84.
65. P. Tallet, *Les papyrus de la Mer Rouge 'Journal de Merer'*, Cairo: Institut Français d'Archéologie Orientale, 2017; P. Tallet, 'Les journaux de bord du règne de Chéops au ouadi el-Jarf (P. Jarf AF): État des lieux', *Bulletin de la Société Française d'Égyptologie* 198 (2017): 8–19; Tallet & Marouard, 'The harbor of Khufu'.
66. Tallet, *Les papyrus*, p. 150.
67. Englund, 'Texts', pp. 70–71; J. Taylor, 'Administrators and scholars: The first scribes', in Crawford (ed.), *The Sumerian World*: 290–304, pp. 293, 297–298.
68. M. Al-Rashid, 'Schoolboy, where are you going? Scribal education in the ancient Mesopotamian tablet house', *Lapham's Quarterly* 14(4) (2022).
69. Englund, 'Texts', pp. 82–106.
70. Leick, *Mesopotamia*, pp. 73–74; Englund, 'Texts', pp. 103–105.
71. R. J. Williams, 'Scribal training in ancient Egypt', *Journal of the American Oriental Society* 92 (1972): 214–221, p. 215.
72. Englund, 'Texts', pp. 99–102; N. Veldhuis, 'How did they learn cuneiform? Tribute/Word List C as an elementary exercise', in P. Michalowski & N. Veldhuis (eds.), *Approaches to Sumerian Literature*, Leiden: Brill, 2006: 181–200.
73. Charvát, *Mesopotamia before History*, pp. 220–221.

74. B. Alster, *Wisdom of Ancient Sumer*, Bethesda, MD: CDL Press, 2005, pp. 22–23.
75. https://etcsl.orinst.ox.ac.uk/section5/tr561.htm translates a composite text using later versions.
76. Taylor, 'Administrators and scholars', p. 298.
77. S. Kramer, *The Sumerians: Their history, culture and character*, Chicago: Chicago University Press, 1963, pp. 165–228.
78. Lichtheim, *Ancient Egyptian Literature*, pp. 58, 61.
79. Charvát, *Mesopotamia before History*, pp. 220–221.
80. C. E. Suter, 'Kings and queens: Representation and reality', in Crawford (ed.), *The Sumerian World*: 201–226, p. 201.
81. Suter, 'Kings and queens', p. 203
82. Suter, 'Kings and queens', p. 214.
83. Karmer, *The Sumerians*, pp. 49, 321.
84. Pollock, *Ancient Mesopotamia*, p. 168; Leick, *Mesopotamia*, pp. 86–89.
85. L. Nigro, 'The two steles of Sargon: Iconology and visual propaganda at the beginning of royal Akkadian relief', *Iraq* 60 (1998): 85–102.
86. Glassner, *The Invention*, p. 201.
87. etcsl.orinst.ox.ac.uk/section2/tr215.htm; J. B. Pritchard (ed.), *The Ancient Near East: An anthology of texts and pictures*, Princeton, NJ: Princeton University Press, 2011, pp. 414–423.
88. M. A. Hoffman, *Egypt before the Pharaohs*, London: Routledge, 1980, p. 296.
89. Lichtheim, *Ancient Egyptian Literature*, pp. 15–28.
90. J. Richards, 'Text and context in late Old Kingdom Egypt: The archaeology and historiography of Weni the Elder', *Journal of the American Research Center in Egypt* 39 (2002): 75–102.
91. Richards, 'Text and context', p. 100.
92. web.archive.org/web/20180810083122/http://reshafim.org.il/ad/egypt/texts/weni.htm.
93. B. J. Kemp, *Ancient Egypt: Anatomy of a civilisation*, 3rd ed., London: Routledge, 2018.
94. Lichtheim, *Ancient Egyptian Literature*, p. 55.
95. S. Langdon, *Sumerian Liturgies and Psalms*, Philadelphia: University Museum, 1919.
96. Leick, *Mesopotamia*, pp. 71–72.
97. J. Andersson, 'Private commemorative inscriptions of the Early Dynastic and Sargonic periods: Some considerations', in Balke and Tsouparopoulou (eds.), *Materiality of Writing*: 47–71; J. Baines, 'Display of magic in Old Kingdom Egypt', in K. K. Szpakowska (ed.), *Through a Glass Darkly: Magic, dreams, and prophecy in ancient Egypt*, Swansea: Classical Press of Wales, 2006: 1–32.
98. Lichtheim, *Ancient Egyptian Literature*, pp. 29–50.
99. Lichtheim, *Ancient Egyptian Literature*, p. 30.
100. A. Annus (ed.), *Divination and interpretation of signs in the ancient world*, Chicago: Oriental Institute of the University of Chicago, 2010, p. 1; J. F. Quack, 'New sources for ancient Egyptian divination', in Szpakowska (ed.), *Through a Glass Darkly*: 175–188.
101. Shaughnessy, 'The beginnings of writing', p. 216

102. Kramer, *The Sumerians*, p. 85.
103. A. G. McDowell, *Village Life in Ancient Egypt: Laundry lists and love songs*, Oxford: Oxford University Press, 1999, pp. 28–32.
104. Cooper, 'Babylonian beginnings', p. 84.
105. T. van den Hout, 'The rise and fall of cuneiform script in Hittite Anatolia', in Woods (ed.), *Visible Language*: 99–108, p. 102.
106. J. Firth, J. Touros, B. Stubbs et al., 'The "online brain": How the Internet may be changing our cognition', *World Psychiatry* 18 (2019): 119–129.
107. S. Greenfield, *Mind Change: How digital technologies are leaving their mark on our brains*, London: Random House, 2015.
108. *Phaedrus*, in J. M. Cooper (ed.), *Plato: Complete works*, Indianapolis, IN: Hackett, 1997, pp. 551–552.
109. Translators' note in Glassner, *The Invention*, pp. xii–xv.
110. A. Macdonald & A. Mazel, 'Challenging "prehistory" in South African archaeology', *South African Archaeological Bulletin* 76 (2021): 91–92.

Chapter 5 Inventing Printing

1. T.-S. Tsien, *Paper and Printing, Science and Civilisation in China*, Vol. 5 Part 1, Cambridge: Cambridge University Press, 1985, pp. 143–146.
2. T. H. Barrett, *The Woman Who Discovered Printing*, New Haven, CT: Yale University Press, 2008; Tsien, *Paper and Printing*, pp. 149–150; Li Zhizhong, 'On the invention of wood blocks for printing in China', in S. M. Allen, L. Zuzao, C. Xiaolan & J. Bos (eds.), *The History and Cultural Heritage of Chinese Calligraphy, Printing and Library Work*, Berlin: De Gruyter Saur, 2010: 35–44. This volume has other useful essays.
3. Barrett, *The Woman Who Discovered Printing*, p. 125.
4. Tsien, *Paper and Printing*, p. 151; https://www.bl.uk/collection-items/diamond-sutra.
5. Tsien, *Paper and Printing*, p. 156.
6. Tsien, *Paper and Printing*, p. 174.
7. Tsien, *Paper and Printing*, p. 190.
8. Tsien, *Paper and Printing*, pp. 197–201.
9. Tsien, *Paper and Printing*, pp. 201–222; M. Bussotti & Q. Han, 'Typography for a modern world? The ways of Chinese movable types', *East Asian Science, Technology, and Medicine* 40 (2016): 9–44.
10. Bussotti & Han, 'Typography', p. 15.
11. Bussotti & Han, 'Typography'.
12. A. Kapr, *Johann Gutenberg: The man and his invention*, London: Scolar Press, 1996, pp. 117–118.
13. Tsien, *Paper and Printing*, pp. 211–220; Bussotti & Han, 'Typography', p. 20.
14. Bussotti & Han, 'Typography', p. 26.
15. Tsien, *Paper and Printing*, p. 325.
16. Bussotti & Han, 'Typography', p. 19.
17. Kapr, *Johann Gutenberg*, p. 113.
18. Tsien, *Paper and Printing*, pp. 327, 330; Kapr, *Johann Gutenberg*, p. 114.

19. Kapr, *Johann Gutenberg*, p. 114.
20. P. Kornicki, 'The Hyakumantō darani and the origins of printing in eighth-century Japan', *International Journal of Asian Studies* 9 (2012): 43–70.
21. Tsien, *Paper and Printing*, p. 338.
22. T. Christiansen, M. Cotte, W. de Nolf et al., 'Insights into the composition of ancient Egyptian red and black inks on papyri achieved by synchrotron-based microanalyses', *Proceedings of the National Academy of Sciences* 2020, 202004534.
23. Tsien, *Paper and Printing*, p. 32.
24. J. Bloom, *Paper before Print: The history and impact of paper in the Islamic world*, New Haven, CT: Yale University Press, 2001, pp. 47–49.
25. O. Da Rold, *Paper in Medieval England: From pulp to fiction*, Cambridge: Cambridge University Press, 2020.
26. Da Rold, *Paper*, pp. 7, 27, 37, 52.
27. R. McKitterick, 'Script and book production', in R. McKitterick (ed.), *Carolingian Culture: Emulations and innovation*, Cambridge: Cambridge University Press, 1994: 221–247, p. 237.
28. L. Febvre & H.-J. Martin, *The Coming of the Book: The impact of printing 1450–1800* (London: NLB, 1976), p. 16.
29. D. E. Booton, *Manuscripts, Market and the Transition to Print in Late Medieval Brittany*, Farnham: Ashgate, 2010, pp. 18, 23.
30. S. Füssel, *Gutenberg and the Impact of Printing*, London: Ashgate, 2003, p. 113.
31. Füssel, *Gutenberg*, pp. 110–111.
32. K.-W. Chow, 'Reinventing Gutenberg: Woodblock and movable-type printing in Europe and China', in S. A. Baron, E. N. Lindquist & E. F. Shevlin (eds.), *Agent of Change: Print culture studies after Elizabeth L. Eisenstein*, Amherst: University of Massachusetts Press, 2007): 169–192, p. 172.
33. Chow, 'Reinventing Gutenberg', p. 183.
34. Tsien, *Paper and Printing*, pp. 306–307.
35. J. Bloom, *Paper before Print: The history and impact of paper in the Islamic world*, New Haven, CT: Yale University Press, 2001, pp. 90–91.
36. Tsien, *Paper and Printing*, p. 307.
37. Bussotti & Han, 'Typography', pp. 12–13.
38. Kapr, *Johann Gutenberg*, p. 63.
39. Kapr, *Johann Gutenberg*, p. 129.
40. M. Lyons, *A History of Reading and Writing in the Western World*, Houndsmills: Palgrave Macmillan, 2010, p. 28.
41. Kapr, *Johann Gutenberg*, p. 148.
42. Kapr, *Johann Gutenberg*, pp. 189–197.
43. Kapr, *Johann Gutenberg*, p. 202–208.
44. Kapr, *Johann Gutenberg*, p. 246.
45. Kapr, *Johann Gutenberg*, p. 274–276.
46. Füssel, *Gutenberg*, p. 8; Febvre, *The Coming of the Book*, p. 248.
47. A. Pettegree, 'Publishing in print: Technology and trade', in E. Cameron (ed.), *New Cambridge History of the Bible, Volume 3, from 1450 to 1750* (Cambridge: Cambridge University Press, 2016): 159–186, p. 165; A. Pettegree, *The Book in the Renaissance*, New Haven, CT: Yale University Press, 2010, pp. 71–72.

48. Febvre, *The Coming of the Book*, p. 267; E. Buringh & J. Van Zanden. 'Charting the "Rise of the West": Manuscripts and printed books in Europe, a long-term perspective from the sixth through eighteenth centuries', *Journal of Economic History* 69 (2009): 409–445, p. 417.
49. www.ustc.ac.uk/about.
50. Kapr, *Johann Gutenberg*, p. 280.
51. A. Pettegree, *The Book in the Renaissance*, p. 357.
52. J. Verger, 'Schools and universities', in C. Allmand (ed.), *The New Cambridge Medieval History Vol.* VII: *c. 1415–c. 1500*, Cambridge: Cambridge University Press, 1998: 220–242, pp. 233–234; Febvre, *The Coming of the Book*, pp. 250–251.
53. M. Black, *Cambridge University Press 1584–1984*, Cambridge: Cambridge University Press, 1984, pp. 23–24, 40.
54. Füssel, *Gutenberg*, p. 68.
55. J. L. Flood, '"Volentes sibi comparare infrascriptos libros impressos ... ": Printed books as a commercial commodity in the fifteenth century', in J. K. Jensen (ed.), *Incunabula and Their Readers*, London: British Library, 2003, 139–151, p. 140.
56. M. K. Duggan, 'Reading liturgical books', in Jensen (ed.), *Incunabula*: 71–84, p. 72.
57. C. Dondi, 'Books of Hours: The development of the texts in printed form', in Jensen (ed.), *Incunabula*: 53–70.
58. Füssel, *Gutenberg*, p. 56.
59. Füssel, *Gutenberg*, p. 153.
60. Kapr, *Johann Gutenberg*, p. 170.
61. Lyons, *History*, pp. 68–70.
62. Pettegree, 'Publishing in print', p. 165.
63. Febvre, *The Coming of the Book*, p. 248.
64. L. Hellinga, *Incunabula in Transit: People and trade*, Leiden: Brill, 2018, pp. 21–23.
65. Febvre, *The Coming of the Book*, pp. 222–224.
66. M. Vali, 'Manuscripts and books', in Allmand (ed.), *The New Cambridge Medieval History Vol.* VII: 278–286, p. 281.
67. Hellinga, *Incunabula*, pp. 6–19.
68. Febvre, *The Coming of the Book*, pp. 237–239.
69. Tsien, *Paper and Printing*, p. 9.
70. Duggan, 'Reading liturgical books', pp. 72–74.
71. K. Jensen, 'Printing the Bible in the fifteenth century', in Jensen (ed.), *Incunabula*: 115–138, p. 138.
72. H.-J. Martin, *The History and Power of Writing*, Chicago: University of Chicago Press, 1994, p. 347, cited by Lyons, *History*, p. 35.
73. Febvre, *The Coming of the Book*, p. 25.
74. Pettegree, *The Book in the Renaissance*, p. 136.
75. Lyons, *History*, p. 27.
76. Lyons, *History*, p. 44.
77. Febvre, *The Coming of the Book*, pp. 217–218.
78. Febvre, *The Coming of the Book*, p. 249.
79. Pettegree, *The Book in the Renaissance*, p. 357.

80. Pettegree, *The Book in the Renaissance*, p. 158.
81. E. Eisenstein, *The Printing Revolution in Early Modern Europe*, Cambridge: Cambridge University Press, 1983, pp. 161–163.
82. Febvre, *The Coming of the Book*, p. 249, 264.
83. Pettegree, *The Book in the Renaissance*, p. 132.
84. Pettegree, *The Book in the Renaissance*, pp. 290–291.
85. S. H. Hendrix, *Martin Luther: Visionary reformer*, New Haven, CT: Yale University Press, 2015, p. 199; Pettegree, *The Book in the Renaissance*, pp. 172–176.
86. A. Blair, 'Student manuscripts and the textbook', in E. Campi, S. de Angelis, A.-S. Goeing & A. T. Grafton (eds.), *Scholarly Knowledge: textbooks in Early Modern Europe*, Geneva: Librairie Droz, 2008: 39–74, p. 48.
87. J. C. Moore, *A Brief History of Universities*, Cham: Springer, 2019, pp. 21–22.
88. R. Black, *Humanism and Education in Medieval and Renaissance Italy*, Cambridge: Cambridge University Press, 2001, pp. 25, 84.
89. Eisenstein, *The Printing Revolution*, p. 399.
90. Pettegree, *The Book in the Renaissance*, p. 119.
91. A. T. Grafton, 'Textbooks and the disciplines', in Campi et al. (eds.), *Scholarly Knowledge*: 11–35, p. 26.
92. Grafton, 'Textbooks', p. 18.
93. F. Watson, *The English Grammar Schools to 1660: Their curriculum and practice*, Cambridge: Cambridge University Press, 1908, p. 54.
94. Watson, *English Grammar Schools*, p. 3.
95. Pettegree, *The Book in the Renaissance*, pp. 177–199.
96. Watson, *English Grammar Schools*, pp. 232–234.
97. Watson, *English Grammar Schools*, pp. 501–513.
98. A. Grafton & L. Jardine, *From Humanism to the Humanities: Education in the liberal arts in fifteenth and sixteenth century Europe*, Cambridge, MA: Harvard University Press, 1986, p. 24.
99. Grafton & Jardine, *From Humanism to the Humanities*, pp. 103–104.
100. Pettegree, *The Book in the Renaissance*, p. 60.
101. Füssel, *Gutenberg*, p. 106.
102. Pettegree, *The Book in the Renaissance*, p. 83.
103. Füssel, *Gutenberg*, p. 74.
104. Febvre, *The Coming of the Book*, p. 207.
105. J. Poskett, *Horizons: A global history of science*, London: Penguin, 2022, pp. 11–93.
106. R. Derricourt, *Antiquity Imagined: The remarkable legacy of Egypt and the Ancient Near East*, London: I.B. Tauris, 2015, pp. 19–24.
107. Eisenstein, *Printing Revolution*, pp. 140–141.
108. Pettegree, *The Book in the Renaissance*, pp. 130–131.
109. Pettegree, *The Book in the Renaissance*, p. 196; P. Marshall, *Heretics and Believers: A history of the English reformation*, New Haven, CT: Yale University Press, 2017, p. 23.
110. Kapr, *Johann Gutenberg*, p. 46.
111. R. Marsden & E. Matter (eds.), *The New Cambridge History of the Bible Vol 2: From 600 to 1450*, Cambridge: Cambridge University Press, 2012.

112. Füssel, *Gutenberg*, pp. 163, 176.
113. Hendrix, *Martin Luther*, pp. 81, 87, 92.
114. Hendrix, *Martin Luther*, p. 14.
115. Pettegree, *The Book in the Renaissance*, p. 208.
116. Marshall, *Heretics and Believers*, pp. 22–23, 54, 256.
117. Marshall, *Heretics and Believers*, pp. 126, 468.
118. M. Lowry (ed.), *Polemic against Printing*, Birmingham: Hayloft, 1986.
119. Füssel, *Gutenberg*, p. 111.
120. Lyons, *History*, p. 51.
121. A. Johns, 'How to acknowledge a revolution', *American Historical Review* 107 (2002): 106–125, p. 119.
122. Pettegree, *The Book in the Renaissance*, pp. 206–207.
123. Febvre, *The Coming of the Book*, p. 306.
124. Hendrix, *Martin Luther*, pp. 94–95, 98, 181.
125. Johns, 'How to acknowledge a revolution'.

Chapter 6 Communicating Wirelessly

1. H. P. Colburn, 'Connectivity and communication in the Achaemenid Empire', *Journal of the Economic and Social History of the Orient* 56 (2013): 29–52, pp. 41–7.
2. A. M. Ramsay, 'The speed of the Roman imperial post', *Journal of Roman Studies* 15 (1925): 60–74; C. W. J. Eliot, 'New evidence for the speed of the Roman imperial post', *Phoenix* 9 (1955): 76–80; W. Scheidel, 'The shape of the Roman world: Modelling imperial connectivity', *Journal of Roman Archaeology* 27 (2014): 7–32; Colburn, 'Connectivity and communication', p. 48; https://orbis.stanford.edu/.
3. H. Shim, 'The postal roads of the Great Khans in Central Asia under the Mongol-Yuan Empire', *Journal of Song-Yuan Studies* 44 (2014): 405–469.
4. Colburn, 'Connectivity and communication', p. 46.
5. E. H. Seland, 'Camels, camel nomadism and the practicalities of Palmyrene caravan trade', *ARAM* 27 (2015): 59–68, p. 48, citing C. P. Grant, *The Syrian Desert*, London: A. & C. Black, 1937, p. 146.
6. S. Arcenas, 'ORBIS and the sea: A model for maritime transportation under the Roman Empire', online at https://orbis.stanford.edu/assets/Arcenas_ORBISandSea.pdf; https://orbis.stanford.edu/#seatransport.
7. D. Headrick, 'A double-edged sword: Communications and imperial control in British India', *Historical Social Research* 35 (2010): 51–65, p. 54.
8. T. Standage, *The Victorian Internet: The remarkable story of the telegraph and the nineteenth century's on-line pioneers*, London: Walker, 1998, p. 16.
9. B. Winston, *Media Technology and Society: A History. From the telegraph to the Internet*, London: Routledge, 1998, pp. 21–22.
10. A. S. Clifton-Morekis, 'Front-line fowl: Messenger pigeons as communications technology in the U.S. army', *History and Technology* 37 (2021): 203–246.
11. Clifton-Morekis, 'Front-line fowl', p. 206.
12. Clifton-Morekis, 'Front-line fowl', p. 209.

13. Clifton-Morekis, 'Front-line fowl', p. 214.
14. Clifton-Morekis, 'Front-line fowl', p. 129.
15. Winston, *Media Technology and Society*, pp. 23–26; Standage, *The Victorian Internet*, pp. 17–18.
16. Winston, *Media Technology and Society*, p. 26.
17. K. Beauchamp, *History of Telegraphy*, Stevenage: Institution of Electrical Engineers, 2001, pp. 103–108.
18. Beauchamp, *History of Telegraphy*, p. 108.
19. J. Schwoch, *Wired into Nature: The telegraph and the North American frontier*, Urbana: University of Illinois Press, 2018; Beauchamp, *History of Telegraphy*, pp. 110–115.
20. Beauchamp, *History of Telegraphy*, pp. 119–120.
21. Headrick, 'A double-edged sword', pp. 52–53.
22. M. Gorman, 'Sir William O'Shaughnessy, Lord Dalhousie, and the establishment of the telegraph system in India', *Technology and Culture* 12 (1971): 581–601, pp. 584–585, 589–590, 597.
23. Beauchamp, *History of Telegraphy*, pp. 108–110.
24. N. Lahiri, 'Commemorating and remembering 1857: The revolt in Delhi and its afterlife', *World Archaeology* 35 (2003): 35–60, p. 48.
25. Winston, *Media Technology and Society*, pp. 30–50.
26. Winston, *Media Technology and Society*, p. 53.
27. *Historical Statistics of the United States, Colonial Times to 1957*, p. 480, online at www2.census.gov/library/publications/1960/compendia/hist_stats_colonial-1957/hist_stats_colonial-1957-chR.pdf.
28. 'Telephone statistics of the world', *Nature* 134 (1934): 527.
29. *Report from the Select Committee on Lighting by Electricity*, London: H.M. Stationery Office, 1879, p. 69.
30. www.britishtelephones.com/histuk.htm.
31. T. K. Sarkar, R. J. Mailloux, A. A. Oliner, M. Salazar-Palma & D. L. Sengupta, *History of Wireless*, Hoboken, NJ: Wiley, 2006, pp. 1–2.
32. Sarkar et al., *History of Wireless*, p. 29.
33. Sarkar et al., *History of Wireless*, p. 331–332.
34. Sarkar et al., *History of Wireless*, pp. 84–88, 247–266; B. Kendal, 'The beginnings of air radio navigation and communication', *Journal of Navigation* 64 (2011): 157–167, pp. 157–158.
35. Beauchamp, *History of Telegraphy*, p. 308.
36. S. D. Ilcev, 'The development of maritime radio communications', *International Journal of Maritime History* 30 (2018): 536–554, pp. 538–541.
37. A. Marincic, 'Nikola Tesla and his contributions to radio development', in Sarkar et al., *History of Wireless*: 267–289.
38. J. S. Belrose, 'The development of wireless telegraphy and telephony, and pioneering attempts to achieve transatlantic wireless communications', in Sarkar et al., *History of Wireless*: 349–420, pp. 402, 409.
39. Beauchamp, *History of Telegraphy*, p. 315
40. D. Juniper, 'The First World War and radio development', *RUSI Journal* 148 (2003): 84–89, p. 84; N. Arceneaux, 'The Wireless Press and the Great War: An

intersection of print and electronic media, 1914–1921', *Journal of Radio & Audio Media* 26 (2019): 318–335, p. 320.
41. Winston, *Media Technology and Society*, p. 78.
42. Winston, *Media Technology and Society*, pp. 95–98, 112, 125.
43. P. Satia, 'War, wireless, and empire: Marconi and the British warfare state, 1896–1903', *Technology and Culture* 51 (2010): 829–853, pp. 830, 849.
44. A. P. Morgan, *Wireless Telegraphy and Telephony: A practical treatise*, New York: Norman W. Henley, 1912.
45. C. I. Hamilton, 'Naval power and diplomacy in the nineteenth century', *Journal of Strategic Studies* 3 (1980): 74–88.
46. Winston, *Media Technology and Society*, p. 71.
47. Beauchamp, *History of Telegraphy*, p. 310; R. W. Burns, *Communications: An international history of the formative years*, London: Institution of Electrical Engineers, 2004, pp. 350, 401.
48. Ilcev, 'The development', p. 541.
49. A. d'Amico & R. Pittenger, 'A brief history of active sonar', *Aquatic Mammals* 35 (2009): 426–434, p. 426; W. D. Hackmann, 'Sonar research and naval warfare 1914–1954: A case study of a twentieth-century establishment science', *Historical Studies in the Physical and Biological Sciences* 16 (1986): 83–110, pp. 90–99.
50. Juniper, 'The First World War', p. 88.
51. Burns, *Communications*, pp. 120–121, 281–5.
52. Beauchamp, *History of Telegraphy*, p. 250.
53. Ilcev, 'The development', p. 541.
54. Beauchamp, *History of Telegraphy*, pp. 241–242.
55. Beauchamp, *History of Telegraphy*, p. 252.
56. R. May, D. Soroka, W. Presnell & B. Garcia, 'Marine weather forecasting in the National Weather Service', *Marine Technology Society Journal* 49 (2015): 37–48, p. 38.
57. E. B. Calvert, 'History of radio in relation to the work of the Weather Bureau', *Monthly Weather Review* 51 (1923): 1–2.
58. www.antiquewireless.org/wp-content/uploads/50-the_first_wireless_time_signals_to_ships_at_sea.pdf; Burns, *Communications*, pp. 120–121.
59. B. N. Hall, 'The British Army and wireless communication, 1896–1918', *War in History* 19 (2012): 290–321, pp. 294–295.
60. Satia, 'War, wireless', p. 836.
61. Beauchamp, *History of Telegraphy*, p. 267; Hall, 'The British Army', p. 294; Sarkar et al., *History of Wireless*, pp. 444–446.
62. Beauchamp, *History of Telegraphy*, pp. 276–280.
63. Juniper, 'The First World War', p. 85; Beauchamp, *History of Telegraphy*, p. 269.
64. Hall, 'The British Army', pp. 306–307, 317.
65. L. S. Lovell, *Russia in the Microphone Age: A history of Soviet Radio, 1919–1970*, Oxford: Oxford University Press, 2015, pp. 19–20.
66. Kendal, 'The beginnings', pp. 159–160.
67. Beauchamp, *History of Telegraphy*, p. 351.
68. Beauchamp, *History of Telegraphy*, p. 258.

69. Kendal, 'The beginnings'.
70. Kendal, 'The beginnings', p. 160; H. G. Schantz, 'On the origins of RF-based location', *2011 IEEE Topical Conference on Wireless Sensors and Sensor Networks*, IEEE, 2011, online at ieeexplore.ieee.org/abstract/document/5725029.
71. Juniper, 'The First World War', pp. 85–86; Beauchamp, *History of Telegraphy*, pp. 348–349.
72. Beauchamp, *History of Telegraphy*, p. 272; Kendal, 'The beginnings', p. 165; P. Judkins, "Sound and fury: Sound and vision in early UK air defence', *History and Technology* 32 (2016): 227–244, p. 228.
73. Beauchamp, *History of Telegraphy*, p. 355.
74. Beauchamp, *History of Telegraphy*, p. 243.
75. W. E. May & L. Holder, *A History of Marine Navigation*, New York: Norton, 1973, pp. 233–235.
76. L. Coe, *Wireless Radio: A brief history*, Jefferson, NC: McFarland, 1966, p. 85.
77. D. W. Watson & H. E. Wright, *Radio Direction Finding*, London: Van Nostrand Reinhold, 1971, pp. 3–5, 8–9; May & Holder, *A History of Marine Navigation*, pp. 222–266.
78. Sarkar, *History of Wireless*, p. 334.
79. May & Holder, *A History of Marine Navigation*.
80. Satia, 'War, wireless', pp. 844–845; Beauchamp, *History of Telegraphy*, pp. 232–233, 236–237.
81. H. J. S. Tworek, 'How not to build a world wireless network: German–British rivalry and visions of global communications in the early twentieth century', *History and Technology* 32 (2016): 178–200, pp. 183–184.
82. M. L. Hadlow, 'Wireless and empire ambition: Wireless telegraphy/telephony and radio broadcasting in the British Solomon Islands Protectorate, South-West Pacific (1914–1947): political, social and developmental perspectives', Ph.D. thesis, University of Queensland 2016 (online at espace.library.uq.edu.au/view/UQ:411422).
83. Hadlow, 'Wireless and empire ambition', p. 84–85.
84. Beauchamp, *History of Telegraphy*, p. 247
85. Arceneaux, 'The Wireless Press'.
86. Arceneaux, 'The Wireless Press', p. 319; www.bbc.com/news/world-europe-42367551; Tworek, 'How not to build'.
87. Tworek, 'How not to build', p. 180.
88. Arceneaux, 'The Wireless Press', pp. 323, 326.
89. Arceneaux, 'The Wireless Press', p. 326.
90. Beauchamp, *History of Telegraphy*, p. 240; D. Read, *The Power of News: The history of Reuters*, Oxford: Oxford University Press, 1999, pp. 215–216.
91. Read, *The Power of News*, p. 223.
92. Juniper, 'The First World War', p. 86.
93. P. Launiainen, *A Brief History of Everything Wireless: How invisible waves have changed the world*, Cham: Springer, 2018, p. 37.
94. Launiainen, *A Brief History*, p. 40
95. Coe, *Wireless Radio*, p. 27.
96. A. Briggs, *The History of Broadcasting in the United Kingdom, Volume 1: The birth of broadcasting*, Oxford: Oxford University Press, 1995, p. 57.

97. R. Henson, *Weather on the Air: A history of broadcast meteorology*, Boston, MA: American Meteorological Society, 2010, pp. 2, 6, 46–7.
98. R. M Wik, 'The USDA and the development of radio in rural America', *Agricultural History* 62 (1988): 177–188.
99. Burns, *Communications*, pp. 435, 438; Briggs, *The History of Broadcasting Vol. 1*, pp. 229–230.
100. Lovell, *Russia in the Microphone Age*, p. 2.
101. Lovell, *Russia in the Microphone Age*, p. 25.
102. A. Briggs, *The History of Broadcasting in the United Kingdom, Volume 2: The Golden Age of wireless*, Oxford: Oxford University Press 1995, pp. 486–487.
103. Briggs, *The History of Broadcasting, Vol. 2*, p. 479.
104. Briggs, *The History of Broadcasting, Vol. 2*, p. 513.
105. Briggs, *The History of Broadcasting, Vol. 2*, p. 562.
106. Sarkar et al., *History of Wireless*, p. 333.
107. Briggs, *The History of Broadcasting, Vol. 2*, p 575; G. Edgerton (ed.), *The Columbia History of American Television*, New York: Columbia University Press, 2007, pp. 8–10.
108. A. Pinkerton, 'Radio and the Raj: Broadcasting in British India (1920–1940)', *Journal of the Royal Asiatic Society* 18 (2008): 167–191.
109. Briggs, *The History of Broadcasting Vol. 1*, pp. 221–222; L. W. Conolly, 'Shaw and BBC English', *The Independent Shavian* 42 (2004): 59–63.
110. H. J. P. Bergmeier & R. E. Lotz, *Hitler's Airwaves: The inside story of Nazi radio broadcasting and propaganda swing*, New Haven, CT: Yale University Press, 1997, pp. 3–4.
111. M. Cormack, 'Minority languages, nationalism and broadcasting: The British and Irish examples', *Nations and Nationalism* 6 (2000): 383–398.
112. Coe, *Wireless Radio*, p. 129.
113. E. L. White & E. C. Denstaedt, 'Police radio communication', *Transactions of the American Institute of Electrical Engineers* 56 (1937): 532–544.
114. Coe, *Wireless Radio*, p. 128; White & Denstaedt, 'Police radio', p. 537.
115. A. Preda, 'Socio-technical agency in financial markets: The case of the stock ticker', *Social Studies of Science* 36 (2006): 753–782.
116. Preda, 'Socio-technical agency', p. 764.
117. Read, *The Power of News*, pp. 220–221.
118. Beauchamp, *History of Telegraphy*, p. 244.
119. Beauchamp, *History of Telegraphy*, p. 245.
120. Morgan, *Wireless Telegraphy and Telephony*, p. 87.
121. Sarkar et al., *History of Wireless*, p. 276.

Chapter 7 Innovation, Progress and Presentism

1. R. Derricourt, *Creating God: The birth and growth of major religions*, Manchester: Manchester University Press, 2021, pp. 139–140, 184–185; Y. Adler, *The Origins of Judaism: An archaeological-historical reappraisal*, New Haven, CT: Yale University Press, 2022.

2. R. Derricourt, *Unearthing Childhood: Young lives in prehistory*, Manchester: Manchester University Press, 2018, pp. 43–44; S. Mithen, *The Singing Neanderthals: The origins of music, language, mind and body*, London: Weidenfeld & Nicolson, 2005 and Cambridge, MA: Harvard University Press, 2006.
3. I. Gilligan, *Climate, Clothing, and Agriculture in Prehistory: Linking evidence, causes, and effects*, Cambridge: Cambridge University Press, 2019.
4. M. Kohn & S. Mithen, 'Handaxes: Products of sexual selection?', *Antiquity* 73 (1999): 518–526.
5. J. Diamond, *Guns, Germs, and Steel: The fates of human societies*, New York: Norton, 1997; Y. N. Harari, *Sapiens: A brief history of humankind*, London: Harvill Secker, 2014; D. Graeber & D. Wengrow, *The Dawn of Everything: A new history of humanity*, London: Penguin & New York: Farrar, Straus and Giroux, 2021; R. L. Kelly, *The Fifth Beginning: What six million years of human history can tell us about our future*, Oakland: University of California Press, 2016; P. Bellwood, *The Five-Million-Year Odyssey: The human journey from ape to agriculture*, Princeton, NJ: Princeton University Press, 2022; P. Frankopan, *The Earth Transformed: An untold history*, London: Bloomsbury, 2023.
6. C. J. Frieman, *An Archaeology of Innovation: Approaching social and technological change in human society*, Manchester: Manchester University Press, 2021, pp. 14, 17–18, 30.
7. M. Price & Y. Jaffe, 'To err is human: Assessing failure and avoiding assumptions', *Antiquity* 97 (2023): 1617–1619.
8. F. Fukuyama, *The End of History and the Last Man*, New York: Free Press, 1992; F. Fukuyama, *Political Order and Political Decay: From the industrial revolution to the globalization of democracy*, New York: Farrar, Straus & Giroux, 2014.
9. R. Derricourt, 'Pseudoarchaeology: The concept and its limitations', *Antiquity* 86 (2012): 524–531.
10. P. J. Crutzen, 'Geology of mankind', *Nature* 415 (2002): 23; C. N. Waters & S. D. Turner, 'Defining the onset of the Anthropocene', *Science* 378 (2022): 706–708.
11. www.geosociety.org/GSA/GSA/timescale/home.aspx; quaternary.stratigraphy.org/major-divisions.
12. A. Witze, 'Geologists reject the Anthropocene as Earth's new epoch – after 15 years of debate', *Nature* 627 (2024): 249–250.

Further Reading

Chapter 2 Taming Fire

Our understanding of the early uses of fire in human prehistory keeps developing, both with new discoveries and with new interpretations. References to journal articles and book chapters in the notes to the chapter provide the elements of these debates, and the data on which they are based.

Stephen J. Pyne has written what is effectively a long essay on fire in natural and different human contexts: *Fire: A brief history*, Seattle WA: University of Washington Press, 2001.

The argument for fire as an essential for early *Homo* was presented by Richard Wrangham, *Catching Fire: How cooking made us human*, New York: Basic Books, 2009. Even without his argument for early control of fire, this book provides a good description of the benefits of cooked food to human biology.

Still useful as a survey of aspects of modern-era foragers is Richard B. Lee & Irven DeVore (eds.), *Man the Hunter*, Chicago: Aldine, 1968.

A pioneering description of Australian Aboriginal use of fire to manage the land and its resources is Sylvia J. Hallam, *Fire and Hearth: A study of Aboriginal usage and European usurpation in south-western Australia*, Canberra: Australian Institute of Aboriginal Studies, 1979.

A special issue of the journal *Current Anthropology* 58: S16 (2017) was dedicated to articles discussing the early use of fire, developed from papers delivered at a 2015 conference. The editors Dennis M. Sandgathe and Francesco Berna present there a summary of the debates, and individual papers provide valuable case studies.

A broader consideration of 'The interaction of fire and mankind' took place at a 2015 meeting of the British Royal Society, with resultant papers published in the *Philosophical Transactions of the Royal Society Series B: Biological Sciences* 317, issue 1696 (2016); including J. A. J. Gowlett, 'The discovery of fire by humans: A long and convoluted process', and a broad-ranging essay by Stephen J. Pyne, 'Fire in the mind: Changing understandings of fire in Western civilization'.

Chapter 3 Domesticating Horses

New archaeological discoveries and new investigations and interpretations of existing data (including DNA studies) make the early history of horse domestication an area of lively debate. The earlier the period, the greater is the diversity of opinion. Journal articles and book chapters referred to in the notes amplify these issues. Some book-length studies provide more accessible revies of some of the histories of the horse and wheeled vehicles.

Pita Kelekna, *The Horse in Human History*, New York: Cambridge University Press, 2009, is an ambitious and broad ranging book, effectively a world history of the role of the horse.

Another account for a general audience is Sandra L. Olsen (ed.), *Horses through Time*, Boulder, CO: Rinehart, 1996, 2004.

A well-illustrated survey of the regions where horse domestication began and nomadic mounted warriors developed their power is Barry Cunliffe, *By Steppe, Desert and Ocean*, Oxford: Oxford University Press, 2015.

A survey of the archaeological background to the western steppe and beyond, and the links the author proposes to language, culture and horse domestication, is David W. Anthony, *The Horse, the Wheel, and Language: How Bronze-Age riders from the Eurasian steppes shaped the modern world*, Princeton NJ: Princeton University Press, 2010.

An earlier accessible account is Stuart Piggott, *Wagon, Chariot and Carriage: Symbol and status in the history of transport*, London: Thames & Hudson, 1992, although information has developed and dating has changed since that volume.

For the Middle East in particular, a classic work is M. A. Littauer and J. H. Crouwel, *Wheeled Vehicles and Ridden Animals in the Ancient Near East*, Leiden: Brill, 1979.

Robert Drews, *Early Riders: The beginnings of mounted warfare in Asia and Europe*, London: Routledge 2004, is an approachable and valuable discussion of the evidence for horses before and during its military uses.

Peter Mitchell, *Horse Nations: The worldwide impact of the horse on Indigenous societies post-1492*, Oxford: Oxford University Press, 2015, is a perceptive review of later impacts.

Several edited volumes, some derived from specialised conferences, are useful in bringing together different studies and perspectives. Marsha Levine, Colin Renfrew and Katie Boyle (eds.), *Prehistoric Steppe Adaptation and the Horse*, Cambridge: McDonald Institute, 2003, is based on a conference held in 2000, as is S. L. Olsen et al., *Horses and Humans: The evolution of human–equine relationships*, Oxford: Archaeopress, 2006. An earlier related volume is M. Levine et al. (eds.), *Late Prehistoric Exploitation of the Eurasian Steppe*, Cambridge: McDonald Institute, 1999.

See also D. Mills & S. W. McDonnell (eds.), *The Domestic Horse: The origins, development and management of its behaviour*, Cambridge: Cambridge University Press, 2005, and papers in in Philipp W. Stockhammer & Joseph Maran (eds.), *Appropriating Innovation: Entangled knowledge in Eurasia, 5000–1500 BC*, Oxford: Oxbow, 2017.

Useful papers based on a 2010 conference were published as Peter Raulwing, Katheryn M. Linduff & Joost H. Crouwel (eds.), *Equids and Wheeled Vehicles in the Ancient World*, Oxford: BAR, 2019.

Chapter 4 Developing Writing

The history of early writing, particularly that of the Middle East, is less contested than for the developments in fire, horses and wheeled vehicles; the notes indicate some of the key material. There are some useful book-length surveys of early writing, the roles it played and the cultural context in early civilisations, including some classic older works.

Christopher Woods (ed.), *Visible Language: Inventions of writing in the ancient Middle East and beyond*, Chicago: Oriental Institute of the University of Chicago, 2010, is a series of essays and, being based on a museum exhibition, well illustrated.

Stephen D Houston (ed.), *The First Writing: Script invention as history and process*, Cambridge: Cambridge University Press, 2004, is a valuable collection of contributions by different authors.

Karen Radner, Nadine Moeller & D. T. Potts (eds.), *The Oxford History of the Ancient Near East: Volume I: From the Beginnings to Egypt and the Dynasty of Akkad*, Oxford: Oxford University Press, 2020, provides reference background. More specialist is Karen Radner & Eleanor Robson (eds.), *The Oxford Handbook of Cuneiform Culture*, Oxford: Oxford University Press, 2011.

There are many general surveys and discussions of early Mesopotamian and Sumerian culture. Still engaging is the classic work by Samuel Noah Kramer, *The Sumerians: Their history, culture and character*, Chicago: Chicago University Press, 1963. A good account is Gwendolyn Leick, *Mesopotamia: The invention of the city*, London: Penguin, 2001. See also Susan Pollock, *Ancient Mesopotamia: The Eden that never was*, Cambridge: Cambridge University Press, 1999.

Petr Charvát, *Mesopotamia before History*, London: Routledge, 2002, is a thorough account and interpretation of major sites. Jean-Jacques Glassner, *The Invention of Cuneiform: Writing in Sumer*, Baltimore, MD: Johns Hopkins University Press, 2003, is an English translation of an important 2000 French title. The contributors to Harriet Crawford (ed.), *The Sumerian World*, London: Routledge 2012, provide valuable overviews.

Relating to Egypt, see Ilona Regulski, 'The origins and early development of writing in Egypt', *Oxford Handbooks Online* (2015). John Baines, *Visual and Written Culture in Ancient Egypt*, Oxford: Oxford University Press, 2007, collects and updates papers published over a period.

A useful collection of translations, although mainly covering later periods than the era discussed in this volume, is James B. Pritchard (ed.), *The Ancient Near East: An anthology of texts and pictures*, Princeton, NJ: Princeton University Press, 2011.

Alongside formal scholarly editions and translations of early Mesopotamian texts, a convenient, accessible source is https://sumerianshakespeare.com/. A more formal source of 1,798 Sumerian texts is at http://oracc.museum.upenn.edu/etcsri/.

Miriam Lichtheim, *Ancient Egyptian Literature. Volume I, The Old and Middle Kingdoms*, Berkeley: University of California Press, 2006, provides an anthology of Egyptian material.

Chapter 5 Inventing Printing

Whereas chapters of this book based on archaeological research are encountering regular new discoveries, most of the discussion of early printing is based on established sources. Debates are therefore more about the significance and impact of the new technology than the stages of its emergence. Since the history of early European printing is well established, older books continue to have value in discussing the context and impact of the innovation.

A detailed survey of relevant developments in China is by Tsien Tsuen-Hsuin, *Paper and Printing*, as Volume 5, Part 1 of the epic series *Science and Civilisation in China* founded and largely written by Joseph Needham (Cambridge: Cambridge University Press, 1985). An accessible short account of 7th-century innovations in China in the context of printing development is T. H. Barrett, *The Woman Who Discovered Printing*, New Haven, CT: Yale University Press, 2008.

On paper, Jonathan Bloom, *Paper before Print: The history and impact of paper in the Islamic world*, New Haven. CT: Yale University Press, 2001, and Orietta Da Rold, *Paper in Medieval England: From pulp to fiction*, Cambridge: Cambridge University Press, 2020.

The life of Johann Gutenberg is investigated in detail in Albert Kapr (translated by Douglas Martin from the 1986 original), *Johann Gutenberg: The man and his invention*, London: Scolar Press, 1996. An outline for a general audience is John Man, *The Gutenberg Revolution: The story of a genius and an invention that changed the world*, London: Review, 2002.

A shorter, illustrated popular history of printing and its impact is Stephan Füssel (translated by Douglas Martin from the 1999 original), *Gutenberg and the Impact of Printing*, London: Ashgate, 2003.

The British Library Incunabula Catalogue at https://data.cerl.org/istc/ gives an opportunity to see the breadth of products and printers in the first fifty years of printing. On the market for early printed materials, see Kristian Jensen (ed.), *Incunabula and Their Readers*, London: British Library, 2003, and also Lotte Hellinga, *Incunabula in Transit: People and trade*, Leiden: Brill, 2018.

An exceptional classic work is Lucien Febvre and Henri-Jean Martin, *The Coming of the Book: The impact of printing 1450–1800*, London: NLB, 1976, translated from the authors' 1958 work *L'Apparition du livre*, Paris: Albin Michel.

Elizabeth L. Eisenstein, *The Printing Revolution in Early Modern Europe*, Cambridge: Cambridge University Press, 1983, is an accessible presentation of her two-volume landmark work *The Printing Press as an Agent of Change*, Cambridge: Cambridge University Press, 1979. Martyn Lyons, *A History of Reading and Writing in the Western World*, Houndsmills: Palgrave Macmillan, 2010, is a valuable survey which includes challenges to the importance Eisenstein places on the influence of Gutenberg as an individual and of printing on changing world views.

The importance of the book and the history of its early development are covered in Andrew Pettegree, *The Book in the Renaissance*, New Haven, CT: Yale University

Press, 2010, with much useful detail and many insights which set the development of the book in social context. Adrian Johns, *The Nature of the Book: Print and knowledge in the making*, Chicago: University of Chicago Press, 1998, covers the first centuries of print and the relationship between intellectual developments and the printed book.

Useful surveys of the editions of the Bible in manuscript and printed form are in *The New Cambridge History of the Bible, Volume 2, From 600 to 1450* (edited by Richard Marsden & E. Ann Matter, Cambridge: Cambridge University Press, 2012), and *Volume 3, From 1450 to 1750* (edited by E. Cameron, Cambridge: Cambridge University Press, 2016).

Relevant to the impact of printing on the development of Protestantism, see Scott H. Hendrix, *Martin Luther: Visionary reformer*, New Haven, CT: Yale University Press 2015; and for England, Peter Marshall, *Heretics and Believers: A history of the English Reformation*, New Haven, CT: Yale University Press 2017.

Chapter 6 Communicating Wirelessly

The history of developments in wireless technology is relatively well established, although individual authors give their own emphasis to different pioneers. Early histories and accounts are therefore no less reliable than later ones. A number of books add detail or perspective to the account in this volume.

Tom Standage, *The Victorian Internet: The remarkable story of the telegraph and the nineteenth century's on-line pioneers*, London: Walker, 1998, is a broad, popular account of the events and people associated with the development of the telegraph.

For a regional focus, James Schwoch, *Wired into Nature: The telegraph and the North American frontier*, Urbana: University of Illinois Press, 2018.

Brian Winston, *Media, Technology and Society: A history. From the telegraph to the internet*, London: Routledge, 1998, gives a useful account of scientific developments and their creators.

Ken Beauchamp, *History of Telegraphy*, Stevenage: Institution of Electrical Engineers, 2001, is a detailed survey of the development of cabled and wireless telegraphy in peace and war, on land and at sea, through to modern times.

A brief, older popular history is Lewis Coe, *Wireless Radio: A brief history*, Jefferson, NC: McFarland 1966, with an emphasis on the USA and individuals in the development of radio technology.

The contributors to T. K. Sarkar et al., *History of Wireless*, Hoboken, NJ: John Wiley & Sons, 2006, provide details of the stages of development of wireless technology and the thought and experiments that preceded it.

Another recent survey is Vinayak Laxman Patil, *Chronological Developments of Wireless Radio Systems before World War II*, Singapore: Springer, 2021.

Russell W. Burns, *Communications: An international history of the formative years*, London: Institution of Electrical Engineers, 2004, is a useful, authoritative and broad-ranging survey.

Further Reading

Asa Briggs, *The History of Broadcasting in the United Kingdom, Volume 1: The birth of broadcasting* and *Volume 2: The golden age of wireless*, Oxford: Oxford University Press, 1995, is a detailed classic account.

On the beginnings of the television era, see Joseph H. Udelson, *The Great Television Race: A history of the American television industry 1925–1941*, Tuscaloosa: University of Alabama Press, 1982.

Index

Aborigines, Australian, 22, 26, 29, 30, 34, 35, 127, 253 n22
Abu Salabih, 120
Abydos, 97, 104–105, 112, 124, 125
Achaemenids, 2, 74, 77, 189, 193
Acheulean, 15, 240–241
Africa, 13, 14, 15, 18, 19, 20, 22, 27, 28, 29, 30, 33, 41, 198, 211; *see also* countries
agriculture, 55, 57, 83, 91, 114, 240, 245
aircraft, 204, 212, 213, 239
Akkadians, 100, 123, 129
Alexanderson, Ernst, 213
alphabetic script, 107
Altyn Tepe, 60
Amarna, 131
Americas, 69, 72, 157, 237, 238; *see also* North America, South America
Anatolia, 56, 64, 107
Anthropocene, 248
Araña, 33
archaeology, 7–8, 13, 46, 54, 134, 243, 244
Ardeles, 38
Armenia, 65
Assyrians, 64, 65, 67, 72–73, 76–77, 82
aurochs, 53
Australia, 19, 22, 34, 35, 37, 40, 74, 79, 199; *see also* Aborigines
Australopithecines, 14
Avicenna, 174, 176

Bacon, Francis, 136
Baghdad, 144
Baird, John Logie, 205, 221
BBC. *See* British Broadcasting Corporation
Belgium, 184, 213
Bell, Alexander Graham, 200
bibles, 146, 152, 153, 156, 163, 173, 175, 180, 183
Blombos, 93
Book of Hours, 146, 156, 182
books. *See* printing
booksellers, 146, 161, 165, 170
Border Cave, 32
Botai, 51–54, 56
Boxgrove, 47
Braun, Karl Ferdinand, 203, 223
Britain, 75, 190, 198, 200, 205, 207, 211, 215, 221, 229; *see also* England
British Broadcasting Corporation, 205, 218, 221–222, 224
Bronocice, 59
Bronze Age, 51, 55, 62, 63, 67, 68, 80, 86
Budakalász, 60
Buddhism, 139, 142
Buhen, 65
Bulguksa Kyongju, 142
Bushmen. *See* San

286

Index

cable. *See* telegraphy
Calahan, Edward A, 227
Calvin, John, 179, 182
Cambridge University Press, 155
camels, 82, 191
Canada, 210
carriages, 69, 75, 84–85, 190
carts, 56–61, 83; *see also* wagons
Cascajel, 110
Catholicism, 70, 146, 151, 155, 158, 162, 166, 179–183
Caucasus, 64
cavalry, 77–79
Caxton, William, 155, 167
Central America, 35, 110
Central Asia, 48, 51, 60, 68, 71, 81, 82, 144
Chaco Canyon, 56
Chanhu-daro, 84
Chappe, Claude, 193
chariots, 61–66, 79–81
Charles V, Emperor, 184
Charvát, Petr, 115
Chaucer, Geoffrey, 145, 155, 167
Chauvet, 48
Childe, Vere Gordon, 91
children, 25, 29, 87
China, 65, 66, 68, 77, 79, 81, 84, 86, 108–109, 121, 129, 138–142, 144, 193, 195, 243, 244
Christianity, 145, 239, 246, 247; *see also* Catholicism, Protestantism
Cicero, 173
Cimmerians, 76
civilisation, 7, 8, 90, 91, 134, 244, 245
class, 92, 121, 236
climate change, 3, 249
clothing, 239
coins, 147
colonialism, 198, 207, 215–216, 224, 243
Columbus, Christopher, 169
communication, 74–75, 88, 187–206, 236; *see also* printing, wireless, writing

computers, 239
Constantinople, 174, 192
contracts, 112
cooking. *See* fire
Copernicus, Nicolaus, 165
Copper Age, 52, 54, 55, 56, 60, 61
Counter-Reformation, 183
couriers, 122, 189, 237
COVID, 248
cremation, 40
Crete, 103, 107
crime. *See* police
Crippen, Hawley, 226
crystal set. *See* wireless
cuneiform, 100–102, 107
cylinder seals, 95, 102
Czechia, 26, 39

Dante Alighieri, 167
Darwin, Charles, 10, 42, 249
Davy, Edward, 196
de Mendoza, Juan Gonzales, 148
Deir el-Medina, 131
Dereivka, 55
dharani, 139, 143
discovery. *See* innovation
divination, 108, 129
DNA. *See* genetics
Dolet, Etienne, 184
Dolní Věstonice, 39
domestication, 45, 51, 237; *see also* horses
Donatus, 151
donkeys, 47, 56, 57, 58, 61, 64, 82
Drimolen, 14
Dunhuang, 139

Easter Island, 108
ebooks, 136, 236
Edelcrantz, Abraham, 194
Edison, Thomas, 227

Egypt, 29, 61, 64–66, 74, 80, 82, 92, 97, 102–106, 113, 114, 115, 117, 120, 124, 128, 130, 148
Einhornhöhle, 40
electricity, 12, 43, 201, 239
Eltville, 154
England, 47, 57, 82, 83, 86, 155, 163, 166, 182, 197, 203; *see also* Britain
Epic of Gilgamesh, 120
Erasmus, Desiderius, 171, 174, 175
Erfurt, 180
Europe, 15, 17, 18, 22, 27, 32, 48, 52, 55, 58, 60, 63, 67, 70, 75, 77, 79, 81, 82, 84, 86, 144, 154, 190, 192, 224, 243; *see also* countries
Evron Quarry, 15

fabric, 139, 149
Fara, 100, 118, 121, 129
Fessenden, Reginald, 203, 208, 210, 213
Fiji, 216
Filippo de Strata, 183
fire, 10–43
　cooking, 17, 29–32
　cooking hypothesis, 16, 32
　firestick farming, 34
　hearths, 13, 15, 19, 20, 24, 26–28, 30
　heat, 27–28
　Homo sapiens and fire, 18–20
　light, 23–24, 26, 38
　lightning, 14, 18
　methods, 20–22
　myths, 41
　natural fire, 12, 13–16
　Neanderthals and fire, 16–18
　signalling, 24–26
Flintbek, 39, 58, 59
Florence, 161
foragers, 14–22, 23–41, 70; *see also* Palaeolithic
Foxe, John, 183

France, 17, 20, 24, 38, 48, 71, 145, 154, 155, 163, 171, 193, 197, 198, 200, 210, 223
Frazer, Sir James, 41
Frederick II, 145
Fukuyama, Francis, 244
Funnel Beaker period, 59

Gabet, Gustav, 208
gas, 12, 43
genetics, 17, 50, 51, 56, 73
Germany, 40, 47, 49, 58, 58, 59, 70, 145, 148, 150–154, 155, 179, 208, 211, 213, 214, 215, 217, 225
Gesher Benot Ya'aqov, 15
Göbekli Tepe, 238
Goebbels, Paul Josef, 225
Graeber, David, 114
Greece, 65, 67, 81, 83, 84, 86
Greenfield, Susan, 132
Gutenberg, Johann, 151–154, 157

Hadza, 29, 30
Halaf period, 94
handaxe, 14, 240
Harappa, 84
Hertz, Heinrich Rudolf, 201
hieratic, 104
hieroglyphs, 102–106
Hittites, 65, 107, 131
Holland. *See* Netherlands
Homo erectus, 14, 15, 16, 32, 240
Homo genus, 14, 249
Homo habilis, 16
Homo heidelbergensis, 47
Homo neanderthalensis. *See* Neanderthals
Homo sapiens, 12, 18, 28, 32, 240, 249
horses, 44–88, 189
　biology, 47, 48, 49–51, 63
　cheek pieces, 55, 62, 63, 259 n35
　domestication, 49–56
　haulage, 81–84

Index

leather, 53, 71
meat, 70–71
milk, 53, 60, 68, 71
Przewalski horse, 48, 51
riding, 52, 60, 66–69, 72–79, 84–86
Hülsmeyer, Christian, 214
humanism, 173–177
Humboldt, 110
Hungary, 60, 64
hunter-gatherers. *See* foragers
hunting, 24, 33, 41, 47, 52, 72–73
Hyksos, 64, 65

Ice Age, 17, 32, 48, 49, 50, 70
Inca, 82, 110, 189
incunabula, 154, 168
India, 68, 79, 81, 192, 193, 198, 199, 215, 224
Indo-European languages, 60, 68
indulgences, 155
Indus, 57, 107
Innocent VIII, Pope, 183
innovation, 1–4, 7, 89, 111, 131, 184, 206, 236–242
Instructions of Shuruppak, 120, 130
internet, 132, 177, 231, 236
invention. *See* innovation
Iran, 59, 61, 76, 98, 107
Iraq, 61, 64, 72; *see also* Mesopotamia
Ireland, 225
Iron Age, 68, 77
Islam, 70, 81, 145, 148, 174, 239, 246
Israel, 15, 17, 37
Italy, 58, 83, 144, 154, 155, 167; *see also* Romans

Jackson, Henry, 202
Japan, 69, 143, 238, 243
Jebel Irhoud, 18
Judaism, 70, 239, 247

Kamennyi Ambar, 63
Kassites, 65, 87

Kazakhstan, 52, 70, 71; *see also* Botai
Kemp, Barry, 127
Kitchen, Jack, 208
Koobi Fora, 15
Korea, 142
Krasnyi Yar, 53
Krivoe Ozero, 63
Kültepe, 64
kurgan, 60, 63, 80

labels, 104, 112
Lagash, 123, 130
lamps, 23–25
language, 15, 16, 93, 104, 127, 131, 166, 223–225, 239, 244
Lascaux, 24, 48–49
Latin, 118; *see also* printing
Lazaret, 20
Lee, Richard, 28
legal texts, 130
Leo X, Pope, 183
Lévi-Strauss, Claude, 31
lexical lists, 118, 119
Liberia, 108
light, 23–24, 201
literacy, 134, 163, 236
literature, 119–121
London, 155, 184, 196, 221, 227
Luther, Martin, 167, 179, 180, 181, 182

Mainz, 151, 153, 154, 157
Malawi, 33
Mandeville, Travels of Sir John, 178
manuscripts, 142, 145–147
maps, 169
Marconi, Guglielmo, 202, 211, 213, 215, 216, 221; *see also* Wireless Telegraph and Signal Company
Marmes, 40
Maszycka, 26
Maxwell, James Clerk, 201
Maya, 110
Mbuti, 22, 41

measurements. *See* numerals
meat, 17, 29, 30, 31, 34, 47, 48; *see also* horses
Medes, 76
medieval. *See* Middle Ages
Mesopotamia, 56, 57, 58, 59, 61, 64–66, 74, 94, 95–102, 115, 118; *see also* Akkadians, Assyrians, Sumerians
Mexico, 71, 155
Middle Ages, 82, 84, 145
Middle East, 57, 62, 64, 65, 91, 144, 148; *see also* countries
Middle Stone Age, 18, 19, 32, 33, 37
military, 73, 75, 76–79, 82, 195, 197, 198, 199, 204, 210–213, 214, 215
Milton, John, 162
Minoan, 103
money, 148
Mongolia, 67, 68, 70, 71, 74, 86
Mongols, 46, 75, 77–78, 140, 147, 190
Montgomery, Sir Robert, 199
Morse code, 203, 207, 209, 214, 217, 225
Morse, Samuel, 196, 197
motor vehicles, 46
Mungo, 40

Native Americans. *See* United States
navy, 202, 206–208, 210, 215, 229
Neanderthals, 16–18, 32, 40, 47
Near East. *See* Middle East
Neolithic, 59
Nerja, 20
Netherlands, 58, 175, 184
New Guinea, 22
New York, 226, 227
news, 199, 216–218
Ngandong, 14
Nicolas of Cues, 148
Nineveh, 72, 76
Nippur, 101, 124, 128, 130
North America, 28, 46, 47, 69, 82; *see also* United States
Norway, 200
numerals, 95–97, 99

ochre, 38
Old Testament, 41, 119
Olduvai Gorge, 14
onagers, 64
oracle bones, 108–109
Ottoman, 148, 157, 207, 208, 211, 224
oxen, 56, 61, 82

Palaeolithic. *See also* Middle Stone Age
 Lower, 16, 19, 20, 23, 47
 Middle, 16, 19, 32, 42, 47
 Upper, 16, 18, 19, 22, 23, 25, 26, 27, 29, 32, 38, 39, 40, 48, 70
Palladio, Andrea, 176
paper, 143–145, 150, 159
papyrus, 103, 143
parchment, 143
Paris, 169, 171, 184, 193
Parthians, 77, 81
pastoralism, 60, 63, 69, 74, 238
Pazyryk, 68
Peasants War (Germany), 179
Pech-Merle, 48
Persia, 81
Peru, 189
pigeons, 195, 199
Pinnacle Point, 37
Pius I, Pope, 158
Pleistocene. *See* Ice Age
Poland, 26, 59, 79
police, 196, 226
Pony Express, 190, 197
Popov, Alexander, 202, 208
pottery, 238
Předmostí, 26
Preece, William, 200
presentism, 246–250
primates, non-human, 14, 31

Index

printing, 136–186
  authors, 162
  East Asia, 138–145
  educational, 151, 156, 161, 170–173
  English, 155, 167
  format, 159
  French, 171
  German, 181
  Greek, 173, 174, 175
  Hebrew, 148, 167, 175
  illustration, 150, 156, 168, 171, 176
  Latin, 151, 154, 156, 160, 166
  legal, 169
  metal type, 141, 142, 149, 152
  music, 170
  pamphlets, 154, 157, 166, 169, 178, 179
  production, 137, 140, 143, 146, 149, 152, 154, 155, 165, 166, 170
  science, 168, 169, 176
  travel literature, 169, 176
  vernacular, 166–168
  woodblock, 139–142, 147, 168
progress, concept of, 242–246
propaganda, 122–127, 178–179, 216–218
Protestantism, 152, 165, 173, 174, 179–183, 184
Proto-Elamite, 107
publishing. *See* printing
punishment, 29
Pyramid Texts, 129

Qesem, 17, 37

radar, 214
radio. *See* wireless
railways, 46, 75, 84, 196, 198
Ramus, Petrus, 171
Rathenau, Erich, 202
Reformation, 157, 179–183
religion, 38, 127, 240
Reuters, 199, 216, 217, 218, 227
rock paintings, 24, 38, 48–49, 70, 90

Roman Catholic Church. *See* Catholicism
Romans, 51, 67, 81, 83, 84, 190, 191, 192, 195
Russia, 54, 56, 60, 79, 86, 192, 198, 202, 208, 212, 221, 244

Sahara, 65
San, 21, 27, 28, 30, 33, 93
Saqqara, 103
Sarnoff, David, 223
Sasanians, 81
Schoeningen, 47
scribes, 117–121, 145, 158
Scythians, 54, 76
semaphore, 193–194
Shaw, Bernard, 224
ships, 191–193, 209–210
Sibudu, 32, 37
Sibyllenbuch, 151
signals, 24–26
silk, 144
Silk Road, 139, 141, 148
Sintashta, 63
smartphone, 188, 232
Socrates, 133
Solomon Islands, 216
Solutré, 48
South Africa, 32, 37, 93, 192
South America, 47, 82; *see also* countries
South Pacific, 216
Soviet Union, 224; *See also* Russia
Spain, 15, 16, 20, 33, 38, 51, 144, 192, 258 n21
speciesism, 248
Sredny Stog culture, 55
stage coach. *See* carriages
stamp seals, 94
steam power, 238
steppe, 51, 52, 54–56, 57, 62, 67, 71
stone tools. *See* tools
Stonehenge, 57

Strasbourg, 148
Strecker, Karl, 202
Sumerians, 58, 64, 67, 92, 94, 96, 98–102, 112, 116, 120, 123, 129, 130
surplus, 92, 115
Sweden, 194, 200
symbolic behaviour, 38
Syria, 59, 61, 64, 94, 107

telegraphy, 196–200
telephone, 200, 231
television, 205, 221–223
Tell Abu Salabih, 100
Tell al-Rimah, 64
Tell Beydar, 61
Tell Brak, 61
Tell Sabi Abyad, 94
Tesla, Nikola, 187, 203, 215, 230
Thebes, 65, 131
time signal, 225
time/space, 6
Titanic, 209
Tlaltenco, 110
tools, metal, 91, 238
tools, stone, 15, 17, 18, 22, 32, 36, 37, 53, 240–241
tools, wooden, 37
tradition, 243
Trithemius, Johannes, 146
Turkey, 238
Turkmenistan, 60
Tyndale, William, 181, 182
type, metal. *See* printing

Ubaid period, 94
Uffington, 86
Ugarit, 107
Ukraine, 54, 55, 60
United States, 14, 20, 35, 40, 56, 72, 75, 79, 86, 108, 190, 195, 196, 198, 199, 200, 203, 205, 208, 215, 222, 226
universities, 146, 155, 161, 162, 170
Ur, 59, 99

Urals, 54, 62, 63
Urban Revolution, 91, 238
Uruk, 58, 94, 98, 99, 102, 112, 117, 118, 119

Vail, Alfred, 197
Vedic, 40, 41, 68
vellum, 143, 146, 153
Venice, 155
Vesalius, Andreas, 176
Vespucci, Amerigo, 157, 169
Vogelherd, 49–50
Volga-Don, 51, 56, 63

Wadi el-Jarf, 103, 116
Wadi Kubbaniya, 29
wagons, 55, 56–61, 83; *see also* carts
warfare. *See* military
water management, 114, 122
watercraft, 57; *see also* ships
waterways, 83
weapons, 208
weather. *See* wireless
Wengrow, David, 114
wheeled vehicles. *See* carts, chariots, wagons
Wi-Fi. *See* internet
Willandra, 37
wireless, 187–234
    CB radio, 232
    crystal set, 188, 205
    domestic radio, 205, 216, 221, 224, 230
    financial information, 227
    low power devices, 230
    maritime, 209–210
    military, 210–213
    naval communication, 206–208
    navigation, 213–214
    physics, 201–204
    radar, 214
    radio waves, 201–206, 234
    short-wave radio, 226
    weather, 209, 225
Wireless Press, 217

Index

Wireless Telegraph and Signal
    Company, 202, 209, 229
wisdom literature, 119
Wittenberg, 181
Wonderwerk, 15
World War, First, 195, 204, 208, 211,
    213, 215, 217, 229
World War, Second, 195, 208, 215, 222,
    225, 230, 232

Wrangham, Richard, 16
writing, 89–135; *see also* cuneiform,
    hieratic, hieroglyphs

Yamnaya, 55, 60

Zoroastrianism, 40, 119
Zwingli, Huldreich, 179, 182
Zworykin, Vladimir, 223